*Selected Topics in*

# Clinical Bacteriology

*Selected Topics in*
# Clinical Bacteriology

*edited by*
## J. de Louvois M.SC., M.I.Biol., F.I.M.L.S.

*Principal Microbiologist,*
*Queen Charlotte's Hospital*
*(for Women), London*

BAILLIÈRE TINDALL LONDON

A BAILLIÈRE TINDALL book published by
Cassell & Collier Macmillan Publishers Ltd
35 Red Lion Square, London WC1R 4SG
and at Sydney, Auckland, Toronto, Johannesburg
an affiliate of
Macmillan Publishing Co. Inc., New York

First published 1976

ISBN 0 7020 0601 7

Printed by Page Bros (Norwich) Ltd, Norwich.

# Preface

The rate of change in clinical bacteriology, as in all biological sciences, is such that it is impossible for the standard works to be truly up-to-date or adequately to reflect topics of current interest.

In undertaking this book, it has been my intention to collect contributions on a number of aspects of clinical bacteriology in which there has been significant recent progress. We have also endeavoured as far as possible to attain a balance between the laboratory and hospital aspects of clinical bacteriology and to concentrate on topics which are currently of concern to many microbiologists. It is regrettable that for economic reasons, the book is restricted to eight contributions and therefore a number of aspects which are currently receiving considerable attention have had to be omitted.

I have been most fortunate in gathering such eminent authors for this volume and would like to express my appreciation to them for their contributions and for the enthusiasm with which they have continued to send additional new material.

I must also express my appreciation to Mr G. D. Wasley for his help with this venture.

*July 1976*                                                    John de Louvois

# List of Contributors

Mr M. J. BYWATER, *Department of Medical Microbiology, Southmead Hospital, Bristol BS10 5NB, UK.*

Dr G. COLMAN, *Department of Dental Science, Royal College of Surgeons of England, Research Establishment, Downe, Orpington, Kent, UK.*

Dr R. FREEMAN, *Lecturer in Clinical Microbiology, School of Medicine, University of Leeds, Leeds LS2 9NL, UK.*

Mr L. R. HILL, *Deputy Curator, National Collection of Type Cultures, Central Public Health Laboratory, Colindale Avenue, London NW9 5HT, UK.*

Dr DONALD A. LEIGH, *Consultant Microbiologist, Wycombe & Amersham General Hospitals, High Wycombe, Bucks HP11 2TT, UK.*

Dr W. C. NOBLE, *St John's Hospital for Diseases of the Skin, Hammerton Grove, London E9 6BX, UK.*

Dr ALAN K. PIERCE, *Department of Internal Medicine, Southwestern Medical School, 5323 Harry Himes Boulevard, Dallas, Texas, USA.*

Dr D. S. REEVES, *Consultant Microbiologist, Southmead Hospital, Bristol, UK.*

Professor JAY P. SANFORD, *Professor of Internal Medicine, Southwestern Medical School, Department of Internal Medicine, 5323 Harry Himes Boulevard, Dallas, Texas, USA.*

Mr G. SYKES, *Principal Inspector, Medicines Division, Department of Health and Social Services, Finsbury Square House, Finsbury Square, London EC2A 1PP, UK.*

# Contents

# 1 | Infection and Intensive Care

## R. FREEMAN

The bacteriological problems encountered on intensive care units can be related to the types of patients commonly admitted to such units, and to the procedures which each patient undergoes as a consequence of intensive care. A general purpose intensive care unit will, over a period of time, include amongst its patients cases of multiple trauma, open heart surgery, tetanus, myocardial infarction and respiratory failure. In addition to these patients, who require intensive care as a primary element of their treatment, it is usual for many other patients, representing those suffering most severely from many quite common conditions, to be nursed in the intensive care unit although they may not require the specialist facilities which are provided. Finally, the intensive care unit is often the place where rarities are found, especially patients afflicted with obscure neurological diseases necessitating ventilatory support.

From the bacteriologist's point of view all patients can be classified into two groups:

(1). Patients with a known, or presumed, infection at admission (dirty cases).

(2). Patients with no infection at admission (clean cases).

Occasionally a third group must be considered in whom an abnormal propensity to infection clearly exists. Examples of this latter group would be patients on immunosuppressive therapy or patients with a disease affecting the immunological system, for instance leukaemia. Fortunately, these latter patients are rarely encountered since independent specialized facilities usually exist for their care.

The role of the clinical bacteriologist on the intensive care unit is, therefore, four-fold: to assist the efficient treatment of the infected cases, to prevent the acquisition of infection by the uninfected cases, to provide the most rapid diagnosis of infection consistent with accuracy, and to promote research into means of improving the situation of his patients. The latter point is especially important when it is realized that an acquired infection can vitiate the results of open heart surgery or any other of the very expensive and time-consuming modern surgical techniques. As medicine advances, the bacteriologist is constantly presented with new settings within which infection must be prevented and treated, and many of the patients involved will pass through the intensive care unit. The aims of this article are to point out some of the current problems and to show some methods of dealing with a constantly changing challenge. Since an exhaustive account is both impossible, and even undesirable, it has been decided to concentrate on certain aspects which, as well as providing specific examples, will serve to illustrate the approach to bacteriology in the setting of intensive care. For this purpose, experiences gained in the Intensive Care Unit of the General Infirmary, Leeds will be described. Firstly, however, it is necessary to make a few general points.

It is too little appreciated by all doctors, although surprisingly well known to intensive care nurses, that patients on an intensive care unit are frequently devoid of the familiar landmarks of infection. Thus, the physical signs of abdominal sepsis will not, and cannot, be elicited in a patient who is fully paralysed for the purposes of ventilation. In some other patients, usually those with head injuries, the conscious level is so depressed that physical signs may be absent without concomitant paralysis therapy. Another example, is found in those patients receiving intravenous hyperalimentation (intravenous feeding). These patients frequently develop septicaemias, originating from the indwelling catheters through which the intravenous solutions are given, without manifesting any localizing signs pointing to the source of the infection (Ashcraft & Leape, 1970). One group of such patients was described from this unit, in which septicaemia resulted from the prolonged giving of more inocuous fluids by this route and in all these patients little or no abnormality was found at the infusion sites (Freeman & King, 1975d). Possible reasons for this state of affairs have been suggested (Craddock et al., 1974).

The second general point, which follows on from the confusing lack of the usual pointers to infection, is that many patients undergoing intensive care may have an equally confusing plethora of investigational results suggestive of infection, although other possible explanations may also be apparent. For instance, many patients will have an increased white cell count in the peripheral blood and, whilst this might well represent an infection, it is almost the rule in patients with uraemia, or, indeed, in uncomplicated recovery from open heart surgery. Thus, many of the investigational landmarks of infection are unreliable in intensive care patients. Pyrexia, in particular, is completely non-specific in many of these patients.

The other side of this subject has equally perplexing problems. Since intensive care patients are a highly selected group, they are very likely to be receiving specialized forms of treatment. Some of these forms of therapy are peculiar to the intensive care unit, for instance mechanically assisted ventilation via endotracheal tubes, and some are forms of therapy used in any severely ill patient, but which appear on the intensive care unit more commonly than elsewhere. Examples of the latter are intravenous feeding, renal dialysis and continuous catheter drainage of the urinary bladder.

All these procedures involve either the profound alteration of a normal physiological mechanism or the insertion of a foreign substance into the body, and some include both. It can therefore, be expected that the potential for infection as a consequence of these procedures exists and must be both guarded against and swiftly detected.

Finally, it is important to mention the thorny problem of antibiotics. It used to be common for all severely ill patients, and particularly those receiving intensive care, to be given broad spectrum antibiotics as a routine part of their therapy. No doubt in many instances this was well justified as the patient was known to have an infection. However, it is equally certain that in some cases these drugs were being given as a means of prophylaxis. There are several counter-arguments to this practice:

(1). All drugs have side effects, and can occasionally kill. Whilst antibiotics represent one of the safest groups of drugs currently available, it is still true that they can occasionally do considerable harm, and it is tragic that a patient should be harmed by a drug for which no true indication existed (Rosenthal, 1958).

(2). The routine use of common antibiotics may result in emergence of resistant strains and these may populate the ward, resulting in super-infections which may be impossible to control. The excellently documented experience of Price and Sleigh on an intensive care unit at Killearn Hospital should serve as a stern warning to all who still insist on using routine prophylactic antibiotics (Price & Sleigh, 1970).

(3). Short of the disastrous experience of Price and Sleigh, prophylactic antibiotics will tend to fail in preventing infection (this prevention lies in careful techniques and good asepsis) but will promote a state in which, whilst the incidence of infection remains largely unaltered, the type of organism causing the infection will be much more difficult to treat (Dunlop & Murdoch, 1960).

(4) The use of prophylactic antibiotics may lead to the masking of a developing infection which would probably have been dealt with if diagnosed and electively treated. Examples are the tragedies which can result from partial and inadequate treatment of post-operative endocarditis following open heart surgery, leading to a recurrence of the infection after discharge from hospital. The destruction of the valve which results would be minimized if the infection had been diagnosed earlier and carefully chosen, well-monitored therapy given. Another example is the masking of a developing abscess by antibiotics, resulting in the subsequent recurrence of the infection after discontinuation of the drugs.

The foregoing is not meant to discourage the use of antibiotics on intensive care units, they are, after all, one of the most potent weapons in the care of these patients, but it is intended to encourage their proper and judicious use. It will become apparent in the following sections that the indications for antibiotic therapy in these patients can be made soundly and specifically for the treatment of infection. An incidental, but very important, corollary of this policy is the remarkable savings which can be made. It was previously the policy for open heart surgical patients on this unit to receive prophylactic broad spectrum antibiotics for up to four weeks after operation. For the last three to four years this has not been the case, and each patient receives antibiotics for only 48 h. It has been calculated that the saving which resulted was several thousand pounds a year, and these patients account for only a proportion of those affected on the unit as a whole. The argument as to whether even the small amounts of prophylactic antibiotic still used are justified is not within the scope of this article. Suffice to say that a minimum objective has been attained, namely, that the prophylaxis is given in such a way that no harm is likely to result; whether any good results is another matter (Sallam et al., 1970).

From the foregoing it can be concluded that one method of dealing with the bacteriological problems of intensive care is to withhold antibiotics and monitor the patients infective status frequently, by serial cultures of urine, sputum, blood and so on, prescribing appropriate drugs when infection has been shown to have occurred. This approach has one unfortunate pitfall; patients undergoing intensive care often become colonized with organisms in particular sites. The distinction between colonization, which may be a harmless process, and infection, is not easily made. One possible approach is for the microbiologist to routinely look for evidence of an inflammatory response in the specimens which he receives. Whilst this may assist in making the distinction between colonization and infection, difficulties still exist. For instance, most patients will produce some degree of inflammatory response in the trachea if an endotracheal tube is in place. The significance of polymorphs in specimens of bronchial secretion is thus of some doubt. Conversely, urine samples may yield significant numbers of organisms unaccompanied by an inflammatory reaction. Finally, this method fails to cope with the silent infections of intravenous catheters since no specimen will be received.

What is, therefore, desperately needed in these patients, is a method which will indicate the presence of infection in the body independent of physical signs, presence (or absence) of organisms in particular sites and unrelated to inflammatory responses of a non-infective nature. The most likely method fulfilling these various criteria which has been described in recent years is the nitroblue tetrazolium (NBT) test (*Lancet*, 1971).

This test, first described for the purposes of detecting bacterial infection by Park et al. (1968), has had a chequered career. Reports on its efficiency in diagnosing infection have appeared in large numbers in recent years, some confirming the original assertions, others questioning the validity of them, (Segal, 1974; Freeman & King, 1975b). The full extent of this controversy is irrelevant to the circumscribed topic under discussion here, and some of the following sections will reveal that the test contributes materially in the context of intensive care.

The full system which is used on the Intensive Care Unit at the General Infirmary Leeds consists of:

(1). Carrying out a daily nitroblue tetrazolium test (NBT).

(2). Frequent (mainly daily) culture and microscopy of urine, bronchial secretion and other relevant specimens, for instance, wound swabs.

(3). Mouth swabs for culture. These are always taken simultaneously with the bronchial secretion specimens.

(4). Culture of all catheter tips on removal. This includes the tips of bladder (Foley) catheters.

(5). Daily chest X-rays.

(6). Daily white cell counts.

(7). Blood cultures are frequently done on patients with intravenous catheters in situ. The culture is always taken from a separate vein.

All cultures are performed on a range of media and thoroughly examined, in order to assess the frequency of resistant organisms.

What follows now is the application of this system to several of the situations which arise in intensive care.

## Chest Infection

Obviously, one of the prime reasons for admission to an intensive care unit is the specialist respiratory care which can be given. Patients requiring this care can be placed into two categories:

(1). Those patients with established infection in the chest, in whom respiratory failure has supervened. In these patients antibiotic therapy will be under way and the microbiological problem is one of assessing the treatment being given and monitoring its efficacy.

(2). Those patients admitted to the intensive care unit for chest conditions requiring ventilatory support but of a non-infective nature, for instance, chest trauma, severe pulmonary oedema or respiratory paralysis, or, those patients receiving intensive care for any reason and in whom bronchopneumonia may develop as a result of immobilization anaesthesia or general debility. The problem in these patients is twofold, that is, firstly to prevent acquisition of a chest infection, and, secondly, to detect the onset of such infection at the earliest possible moment.

The specimens which will be available upon which to judge the efficacy of on-going therapy for an established chest infection or to detect the onset of chest infection will be: Bronchial secretion and mouth swab, Blood culture, White blood cell count, NBT test, and Chest X-ray.

### *Bronchial Secretion and Mouth Swab*

Despite the greatest care in the taking of bronchial secretion samples it soon becomes obvious to the regular observer that true bronchial secretion is rarely received in the laboratory. Microscopy of sputum and bronchial secretion is a sadly neglected procedure, and remains the only firm guide to the value of the specimen. An early series from the Leeds Unit showed very clearly that cells typical of the upper respiratory tract (squamous cells, ciliated columnar cells) are found in appreciable numbers in the majority of samples (Freeman & King, 1973). Simultaneous culture of the bronchial secretion and mouth swab will yield identical organisms in similar proportions in most of these cases and thus, confirms that the bronchial secretion is, in fact, accumulated upper respiratory tract secretion which has trickled down from above the cuff on the endotracheal tube, probably when the cuff is necessarily deflated for short periods. (see Table 1.1).

**Table 1.1.** *Incidence of squamous epithelial cells and the correlation between organisms cultured from bronchial secretion specimens and simultaneously taken mouth swabs in 100 patients with no evidence of chest infection*

| Findings | Pre-intubation | 24 h after intubation | 48 h after intubation |
|---|---|---|---|
| Appreciable numbers of squamous epithelial cells in brochial secretion | 20 per cent | 35 per cent | 70 per cent |
| Same organisms in mouth swab and bronchial secretion | 24 per cent | 25 per cent | 66 per cent |

Furthermore, it is also found that in many patients undergoing antibiotic therapy, a change in mouth flora quickly occurs in which the normal commensal organisms are supplanted by organisms resistant to the antibiotic being given. This change, producing a mouth flora composed of such organisms as *E. coli*, *Klebsiella* spp., pseudomonads and yeasts, will be reflected in the bronchial secretion cultures since, as has already been demonstrated, they are mostly upper respiratory tract material which has collected in the trachea. Table 1.2 shows that the majority of patients will have these changes well established after 48 h of intubation.

In patients with pneumonia (consolidation on the chest X-ray) this pattern may be disturbed. Thus, in the few cases of true lobar (pneumococcal) pneumonia which we have observed, the bronchial secretion has changed in both cytology and organism content to become pure pus

**Table 1.2.** *Sequential changes in mouth flora of 100 patients following institution of antibiotic therapy and endotracheal intubation*

| Findings | Pre-intubation | 24 h after intubation | 48 h after intubation |
|---|---|---|---|
| Normal flora* only | 100 per cent | 7 per cent | 39 per cent |
| Normal flora* plus resistant organisms | – | 20 per cent | 51 per cent |
| Resistant organisms† only | – | 5 per cent | 10 per cent |

* Normal flora implies presence of Streptococci (α haemolytic) and *Neisseria* spp.
† Resistant organisms comprise *E. coli*. klebsiellae. pseudomonads and yeasts.

containing the pneumococcus. However, in bronchopneumonia the pattern has remained much the same as described earlier, and the distinction between a non-infected chest and bronchopneumonia has been impossible to make using microscopy and culture of bronchial secretion. In particular, Table 1.1 shows data obtained from 100 consecutive open heart surgery patients in whom no evidence of chest infection was found using chest X-rays, blood gas analysis, and repeated clinical examinations. It can be seen that in the majority of these patients, organisms were isolated, and, because all these patients were receiving prophylactic short course antibiotics during the study period, many of the organisms isolated were of the resistant type, such as *E. coli.* We have, therefore, concluded that the result of bronchial secretion culture is of little value in the diagnosis and mangement of chest infections in intensive care patients, and, indeed, it may often be positively misleading. It has been our experience that when the patient is extubated and antibiotics are withheld, that all the floral changes will revert to normal. We certainly do not initiate therapy on the basis of bronchial secretion examination, and if therapy is given following a diagnosis of chest infection on other grounds (see below) the nature and sensitivity patterns of any organisms present in the bronchial secretion cultures are not necessarily taken to be relevant. As mentioned earlier, an exception to this rule occurs in cases of lobar pneumonia when the organisms found in the bronchial secretion appear to reflect the process occurring in the lung.

The examination of bronchial secretion should still be carried out however, since it can occasionally help in differential diagnosis. In one such case, which was diagnosed as pulmonary infarction, the chest X-ray was remarkably clear. Microscopy of the secretion revealed an absence of red cells, numerous squamous cells and plentiful Gram-negative rods. Culture of the bronchial secretion and accompanying mouth swab yielded a heavy growth of *Serratia marcescens*, and the orangey-red diffusible pigment produced by this organism explained the macroscopic appearance of the specimen.

Conversely, several patients have arrived on this unit in whom contusion of the lung has occurred as a result of a road traffic accident. After a few days the bronchial secretion has appeared to become purulent whereas the lung lesion, although persisting, has improved. Microscopy of the bronchial secretion has revealed large numbers of haemosiderin-laden macrophages and free haemosiderin. This appearance can be confirmed by a Prussian blue stain, and the apparently purulent nature of the secretion can then be denied and the suspicion of infection refuted. The practised observer will differentiate this condition on the macroscopic appearances since the colour of the secretion is a distinct golden yellow, unlike the dirty yellow found in true purulence.

Finally, it is well worthwhile examining the secretion with Romanowsky stains since the occasional patient in status asthmaticus is admitted to the intensive care unit, and the demonstration that the purulence of the secretion is due to eosinophils and not polymorphs has twice led to a final diagnosis of acute allergic aspergillosis (confirmed by the Sabouraud plate culture) when the patient was being treated for a presumed bacterial infection.

## Blood Culture

Blood cultures are an underused diagnostic procedure throughout medicine, and this is very true in chest infections. Quite often an organism can be recovered from the blood in a patient with a radiological lung lesion. The specific problems of assessing the significance of organisms isolated from blood cultures are well known and need not be discussed further. One interesting extension of this concept was brought home with some force on one occasion on this unit. The patient, a known case of myasthenia gravis in cholinergic crisis, was admitted with a typical lobar pneumonia. Therapy had already been initiated elsewhere and culture of the bronchial secretion and blood failed to yield the pneumococcus which was undoubtedly the cause of the chest lesion. However, when an intravenous catheter, which had been in place for five days, was removed, culture of the tip produced a healthy growth of pneumococcus. This result was repeated from the tip culture of a separate catheter in another vein. The possibility that the thrombus on catheter tips may capture and harbour organisms circulating at a concentration below that detected by conventional blood culture, is worthy of further investigation.

## White Blood Cell Count

As stated earlier, the white cell count can be a fallacious guide in intensive care patients and great caution should be used in its interpretation. These points apart, serial white cell counts can be of value in following the progress of individual patients, particularly those admitted with an existing pneumonia where the pre-admission white cell count is known. Particularly useful, is the study of the differential count, when a progressive fall in the percentage of neutrophils from a previously elevated value gives a good indication that treatment is succeeding. We have also found the old observation that the presence of eosinophils in the peripheral blood indicates absence of bacterial infection or the adequate control of an established infection to be broadly true (Davidsohn, 1962). Finally, in a patient admitted with an established chest infection and no other obvious disease, the absence of a neutrophil leucocytosis should always arouse suspicion of a nonbacterial cause of the infection. To argue in favour of routine paired serology for *Mycoplasma pneumoniae*, Psittacosis and Q fever in all intensive care patients with chest infection would be a philosophy of perfection, but it should be remembered that *M. pneumoniae* has been strongly suspected of being a common agent in acute exacerbations of chronic bronchitis. Since treatment for this organism must be specifically directed, and may not be included in the routine treatment for bronchopneumonia (e.g., ampicillin) the findings referred to above in the white cell count should arouse suspicion, especially if the chest infection appears to be responding poorly to conventional therapy. Similarly at certain times of the year it becomes quite likely that viral pneumonia may present, and an uptodate knowledge of any local epidemics, particularly of influenza virus, when taken together with atypical findings in the white cell count may result in early diagnosis.

## Nitroblue Tetrazolium Test (NBT)

It has been our experience that this test has proved of the greatest help in all our work on infections in intensive care patients. In the specific context of chest infection, it would find its greatest use in the differentiation of the following groups:

(1). Patients with a chest lesion, but in whom no infection exists i.e., the chest lesion is non-infective.

(2). Patients with active uncontrolled chest infections.

(3). Patients in whom chest infection has been convincingly demonstrated and in whom the therapy being given requires to be assessed as either Effective or Ineffective.

As mentioned earlier, it is not within the scope of this article to discuss the merits and demerits of the many techniques and assessments linked with the NBT test. It is, however, necessary to briefly describe its nature and the significance of the results.

The NBT test is based on the observation that when neutrophils are engaged in active killing of ingestion of live cells (in this context, bacteria) their metabolism undergoes a profound change, and hexose monophosphate shunt (HMP) activity becomes greatly enhanced. This results in the formation of hydrogen peroxide within the cell. This substance then takes part in a complex pathway leading to iodination of the organism, and this latter process is responsible for the death of the ingested cell.

This, or similar processes, have been shown to be active against most bacteria, yeasts and mycoplasma (Freeman & King, 1972b). If the polymorph which is undergoing this metabolic burst is exposed to nitroblue tetrazolium (NBT) dye, the dye enters the cell and instead of hydrogen peroxide being produced the dye is reduced by the same pathway and reduced NBT, or formazan, results. NBT dye is a pale yellow solution which on reduction becomes insoluble, crystalline and blue-black in colour. Thus, the affected cells can be identified microscopically by a blue-black deposit within the cytoplasm. In normal and uninfected patients it has been found that 1–10 per cent of neutrophils show this change, whereas in infected patients the proportion is much higher (Park et al., 1968). The test can thus be used in the context of intensive care to monitor the infective status of each patient on a day-to-day basis, especially since the result can be available within 1 h of blood collection (Freeman et al., 1973a). A positive (greater than 11 per cent) result implies that the patient has an uncontrolled bacterial infection or an infection which is not responding to therapy. Conversely, a negative (less than 11 per cent) result implies that the patient does not have an active infection, or that any infection which has been previously present is either cured or is being adequately treated. Negative results are found in viral infections, and in many other conditions which may

**Table 1.3.** *Values obtained in the nitroblue tetrazolium (NBT) test in various infective conditions in intensive care patients*

| Group and number of patients | Range of NBT score (per cent) | Mean NBT score (per cent) |
|---|---|---|
| Controls (no infection present) 50 patients | 1–12 | 7·1 |
| Chest lesion, no chest infection—10 cases* | 1–10 | 6·6 |
| Chest lesion, infection present—10 cases† | 15–44 | 21·9 |
| Septicaemia—10 cases | 21–64 | 43·6 |
| Urinary tract infection— 10 cases | 12–20 | 15·1 |
| Infected catheters— 10 cases | 12–19 | 14·9 |

\* Includes cases of pulmonary thrombo-embolism.
† Includes cases in which radiologically proven atelectasis has persisted for over 24 h despite bronchial toilet.
‡ Implies that no other source of infection (chest, urine, etc.) was found at the time.
   *Statistical note:* Results from patients with a non-infective chest lesion do not significantly differ from controls. In all other groups there is a statistically significant difference.

simulate infection. Occasionally however non-specific suppression of the dye reduction may occur (Freeman et al., 1973a).

Returning to the specific context of chest infections, it should, therefore, be found that the NBT test will distinguish the three groups of patients referred to above. Table 1.3 shows that this has been our experience. Particularly important, is the finding, confirmed by several other workers, that the test will distinguish pulmonary thrombo-embolism (negative tests) from pneumonia (positive test) (Gordon et al., 1974). Further, it is also important to realize that this test can be used in a serial fashion to follow the progress of the patient (Freeman et al., 1973a), and thus to monitor the advent of chest infections in patients initially free of this complication of intensive care. Much remains to be done, and several problems are still outstanding, particularly the fact that a positive NBT test does not localise the site of infection but merely implies that an infection is present somewhere in the body. However, it does seem that this technique may provide a more rational approach to the diagnosis of infection on intensive care units.

## Chest X-rays

Chest X-rays provide the simplest and most convenient method of assessing the lung fields, in intensive care patients. Because of the techniques of respiratory care used (endotracheal intubation, frequent bronchial toilet, etc.) and the prolonged bed rest enforced on most intensive care patients, it is imperative that a regular examination of the lung fields be carried out. Furthermore, many patients already have a pre-existing lung lesion before admission to intensive care and the evolution of this lesion may be thus studied. Unfortunately, whilst being an excellent tool for the demonstration of a lesion, the chest X-ray is limited in its ability to determine the nature of the lesion. This is true, for instance, in the accurate differentiation of pneumonia and pulmonary infarction, or in detecting the onset of suppurative pneumonia in a long-standing area of atelectasis. It is in just this situation that an intelligent appraisal of the results of the NBT test may allow a decision as to the presence or absence of infection to be made.

To summarize our practice in the diagnosis, and management of chest infection in these patients, the following is a general guide.

*Patients with no chest problems prior to admission.* A policy of frequent monitoring, using serial NBT testing, frequent chest X-rays and examination of bronchial secretion and mouth swabs is used. As stated earlier, apparent changes in the bronchial secretion flora, in the absence of any changes in the chest X-ray or NBT status, are mainly ignored and monitoring is continued.

An elevation in the NBT level to more than 11 per cent initiates a review of all possible sites of infection (urine, chest, wounds, and so on) and the finding of a lesion on the chest X-ray in the absence of infection elsewhere in the body leads to consideration of pneumonia. If the lung lesion is of recent appearance, it is initially assumed that the change in the NBT level may represent infection supervening on an area of retained secretion or atelectasis and vigorous bronchial toilet is given. If no improvement in the radiological appearance or in the NBT status is produced after a few hours, antibiotic therapy is commenced and the other measures continued. The choice of antibiotic is usually unrelated to the resistant organisms found in the bronchial secretion, and is usually made on the assumption that the infection is due to the common agents of bronchopneumonia. The subsequent NBT status is then followed to assess the efficacy of the therapy. Blood cultures are taken in an attempt to isolate the causative organism. In a desperately ill patient very potent and broad-spectrum antibiotics are given to produce an immediate effect and reduce the chances of failure.

The appearance of a radiological lesion in the absence of a change in NBT status to positive, even if associated with a leucocytosis and fever, prompts the two diagnoses of early atelectasis or pulmonary embolism. Fortunately, the radiological appearance may allow differentiation, together with such attendant signs as may be found in pulmonary embolism. If atelectasis is suspected, vigorous bronchial toilet is given and the lesion can usually be reversed within a few hours. Serial NBT tests will allow the detection of supervening infection in the collapsed lobe. Pulmonary embolism will require its specific therapy, and sequential examination of the bronchial secretion may reveal the presence of blood followed by the appearance of haemosiderin-laden macrophages. Occasionally, it is possible to detect fat within the macrophages when the embolus follows trauma to the bone (Fat embolism).

Using the twin criteria of an elevated NBT score and the presence of a radiological lung lesion we have found that the occasions on which antibiotic treatment of pneumonia has been necessary have been very few. The number of occasions on which vigorous bronchial toilet has reversed lung changes without the aid of antibiotics have been many. To give a simple example; the incidence of bronchopneumonia in 300 consecutive open heart surgery cases as defined by the above criteria was 1·6 per cent (5 cases) whereas the incidence of radiological lung lesions was over 50 per cent. Only the 5 cases so defined received specific therapy for pneumonia, and the remaining patients received physiotherapy, bronchial toilet and so on. In all the patients with chest X-ray lesions treated without antibiotics, because of normal NBT levels, the bronchial toilet sufficed to clear the chest X-ray within 24 h, and none of them required antibiotic therapy for chest infections throughout the remainder of their post-operative course.

*Patients admitted to intensive care with a chest lesion.* The chest lesions which come within this category include:
(1). Pre-existing bronchopneumonia
(2). Aspiration of gastro-intestinal contents into the lung
(3). Pulmonary contusion (from trauma)
(4). Atelectasis
(5). Pulmonary embolism

Obviously, the first priority in the case of a patient admitted with a known chest lesion is to assess the patient and confirm the diagnosis. Thus, as before, such patients will have the routine examinations performed. In the case of a patient with previously diagnosed pneumonia in whom the NBT score is normal and the chest X-ray shows signs of improvement, no additional treatment may be necessary, and the antibiotics which the patient is already receiving are continued, the bronchial toilet which the intensive care unit provides being added. If, however, the NBT score is elevated and/or the chest X-ray is showing some deterioration, the current therapy must be reviewed. In such a case blood culture may resolve the difficulty.

In patients where aspiration of gastro-intestinal contents has occurred, it is our experience that two grades of severity exist. In some patients simple aspiration of food, e.g. milk, has occurred, and these patients respond well to treatment as for atelectasis above. Thus, bronchial toilet is given and the advent of infection is monitored, antibiotics being given as necessary. In patients, however, where true aspiration of stomach contents has occurred the result is usually a disastrous chemical and then bacterial pneumonitis. All such patients should receive antibiotics, although the outcome is generally poor.

The patient with pulmonary contusion probably illustrates the use of the described system better than most. There is a radiological lung lesion, and, as has been described earlier, the bronchial secretion may assume a yellowish tinge after a few days, prompting the diagnosis of supervening pneumonia. This suggestion can be refuted by the finding of normal NBT levels and confirmation of the true pathology can be obtained by the finding of haemosiderin-laden macrophages in the bronchial secretion. Antibiotics are witheld and serial chest X-rays will show gradual clearing of the opacity.

Atelectasis is assessed as described earlier and treated with bronchial toilet, but antibiotics are given if the change is well established and the NBT score is elevated. Similarly, patients with previously diagnosed pulmonary embolism will be assessed and specific treatment continued as necessary.

## Intravenous and Intra-arterial Catheters

Modern medical care involves the frequent use of intravascular catheters. On the intensive care unit there is a congregation of severely ill patients, many of whom will have such catheters in place. Common examples are those patients requiring monitoring of the central venous pressure (CVP), open heart surgery patients, who may have 4 or more such catheters in situ, and those patients receiving parenteral nutrition. It must be understood that these catheters are not simple drips: they extend for considerable distances along the vessel and the tip is usually in the great veins. Thrombogenesis can be expected to occur after a short period (Rodman et al., 1974), and the situation thus arises in which an environment favourable to bacteria is created. Several questions of considerable importance to the individual patient are thereby raised:

(1). Do catheter tips become infected?

(2). If infection occurs can it be detected, especially before the catheter is removed?

(3). Is the condition important? Does, for instance, dissemination of the organism take place?

(4). Can it be prevented?

Much of our work on intravascular catheters has been done on open heart surgery patients. These patients are a convenient group in this regard since they are the only large group which undergo elective intensive care, allowing preparations to be made, and they also have large numbers of intravascular catheters inserted.

There is a great temptation when a catheter is inserted for a reason which has then passed, for it to remain in place and be used for subsequent infusions which could well be given by a simple drip in a peripheral vein. Our observations extended over two periods, an initial phase in which catheters were left in situ for as long as had been customary, and a second period in which removal of the catheters at the earliest possible moment was advocated. The mean time in situ of all the catheters in the initial series was 3·9 days, and in these catheters organisms could be recovered in 35 per cent. Also notable was the observation that Gram-negative bacilli and yeasts were numbered among the recovered organisms. When early removal was adopted the mean time in situ was reduced to 1·8 days and the isolation rate fell to 8 per cent. Coincident with this fall in the number of organisms isolated was a change in the type of organism. Gram-negative bacilli and yeasts disappeared, and only Gram-positive organism were found (Staphylococci, Streptococci and aerobic sporing bacilli) (Freeman et al., 1973b). The isolations of aerobic sporing bacilli (ASB) were at first an interesting anomaly, but subsequent statistical analysis on the original series and some further data obtained later indicated that these organisms were almost always contaminants. However, statistical analysis of the catheters which yielded organisms other than ASB clearly showed that they behaved as a population quite different from the sterile catheters (Freeman & King, 1975c), and it seems justifiable to use the term infected catheter.

Using the data collected from a large number of open heart surgery patients a statistical analysis was attempted to define the influence of such factors as site, flushing and handling (Freeman & King, 1975a). No firm evidence was obtained but it was of note that, whilst frequent handling of the infusion route was disadvantageous and tended to produce a higher isolation rate, those catheters which were very rarely handled and through which no infusions were made, were by no means free of infection. The opposing effects of stasis, tending to promote thrombogenesis and thus infection, and handling, promoting access of organisms but diminishing thrombus formation by a flushing effect, remain to be more clearly defined. Site was of little significance, although one common site used by others, and claimed to have a low infection rate, namely the radial artery (Gardner et al., 1974), is outside our experience. Infection rates in catheters inserted in theatre under full aseptic precautions were compared with infection rates from catheters inserted on the intensive care unit, and no significant difference was found.

Another point which remains to be resolved is whether the organisms which are isolated from the catheter tip are organisms from the skin (or from outside the patient in a general sense), or whether they are found in the catheter tip thrombus as a result of trapping of organisms circulating at a level below that detected by blood cultures, their origin being internal (gut, mucous surfaces). The progressive more resistant type of organism isolated as catheters are left in situ for longer and longer, would suggest that ingress of external organisms might occur, although changes in the gut and oral flora quickly occur in most patients, especially those on antibiotics. Similarly, our finding of pneumococci in two catheters in a patient with lobar pneumonia with negative blood cultures (see p 9) would also support an endogenous source.

In all the work done on the catherter tips of open heart surgery patients there is a remarkable absence of *Staphylococcus aureus* among the bacteria isolated. However, great pains are taken in this group of patients to clear the skin, nares and perineum of this organism by pre-operative antiseptic treatments (Freeman et al., 1973b). In contrast, *Staph. aureus* does feature in isolations which we have made from non-open heart surgery cases (Freeman & King 1975d). This finding might support an exogenous origin for some of these organisms. Further work needs to be done on these important sites of infection, since prevention of this complication will depend on an accurate knowledge of the relative importance of these two possible routes.

Two further important points must be made in relation to catheter infections. Firstly, we have documented a small series of patients in whom a characteristic syndrome resulted (Freeman & King, 1975d). Briefly, this consists of the development of a fever and general non-specific illness in a patient with indwelling intravascular catheters in place. Characteristically the white cell count remains normal and the infusion site appears innocent, although very minimal signs of inflammation may appear. These patients are often first diagnosed by blood culture as having a septicaemia, usually due to *Staph. aureus*. If the catheter is removed it is found to be the source of the septicaemia, and subsequent administration of appropriate antibiotics results in rapid recovery. Removal is followed by the development of a neutrophil leucocytosis and localized suppuration at the infusion. We have suggested that the presence of the infected catheter, being composed of foreign material represents a situation analogous to the common infection of Spitz-Holtzer valves in hydrocephalic children, in which the inflammatory response is also remarkably poor (Cohen & Callaghan, 1961), or a situation similar to bacterial endocarditis, in which for a different reason the inflammatory response is compromised (Weinstein & Rubin, 1973). In common with both these latter states, it is probable that two courses of treatment are open to pursuit, either long term antibiotic therapy carefully monitored, or removal of the material bearing the infection which is immured from the body defences. Obviously, in the case of the infected catheter, removal is the rational policy. This finding does, however, lead to the formulation of one of the golden rules of intensive care bacteriology. If the patient is ill, and a routine check of the usual sites of infection is unhelpful, any catheters must be removed and cultured. As emphasized above, the finding of an innocent appearance at the infusion site and a normal white blood cell count are not points against the diagnosis of an infected catheter, indeed, they are typical of it. Again, it is in just this dilemma that the NBT test will add weight to what often is ill-received advice, since it acts as an excellent marker for these infections, and does so in a way which is independent of the conventional white cell count (Freeman & King, 1972a).

Finally, the important recent observations on the possible abnormalities in neutrophil function complicating intravenous hyperalimentation, particularly as a result of phosphate depletion, are worthy of note. The first investigators to report this presented evidence that since the common solutions used in intravenous hyperalimentation ('Aminosol', 'Intralipid', and so on) contained little or no phosphate, the patient became depleted. (Craddock et al., 1974). This, in their experimental model, resulted in a fall in neutrophil ATP and subsequent diminution in the neutrophil bactericidal activity. Repletion of the phosphate resulted in rapid recovery, but infection in the depleted patients was impossible to control by any means until this was achieved. Although unable to measure the neutrophil ATP in the same manner, we have since seen three patients with infection complicating intravenous hyperalimentation in whom the serum phosphate was well below the lower limit of normal for the examining laboratory. In two of these patients improvement occurred coincident with phosphate repletion. It thus seems to be a wise precaution to measure the serum phosphate serially in any patient receiving intravenous hyperalimentation, and this we now do.

## Future Trends

Few people can accurately predict future trends in any major branch of science, and this is especially true of medicine. It seems clear that following the major impetus given to microbiology by the discovery and applications of antibiotics more rapid means of diagnosis are needed if further advances are to be made. In this respect, our application of the NBT test would seem to be a small, but not insignificant step. Limitations are still present even if the NBT test is accepted as being valid in this context, the most important being that it does not localize the site of infection, nor the nature of the infecting organism. A more recent test with great potential, and capable of similar application, is the limulus lysate test for the detection of circulating endotoxin (Levin et al., 1970). As with the NBT test, two conflicting schools of thought have emerged but it seems certain that this test should be assessed for use on intensive care patients.

Other workers are busy improving the conventional means of bacteriological diagnosis. Thus, rapid methods of blood culture have been devised and may come to play a part in the management of severely ill patients. On another front, more rapid methods of monitoring treatment, in particular, rapid antibiotic assays, are being developed.

All these methods may eventually become standard procedures in the management of intensive care patients, but perhaps the most important step has already been taken. This is the recognition that all these patients, for various reasons, are a special group, requiring careful study and the accumulation of experience. The days when patients on an intensive care ward were thought merely to be more severely ill examples of other patients, are over.

## Acknowledgements

It must be obvious, even to the most casual reader, that the work which has been described here cannot be the product of one person's industry. I must, therefore, thank my colleagues on the Intensive Care Unit (Ward 21) at Leeds General Infirmary for all their help and encouragement over many years. Any practising doctor will appreciate that in this regard one's nursing colleagues are especially important when accurate taking of specimens is desirable, and my thanks are extended to them. To Mr M. I. Ionescu, Dr J. J. L. Ablett and Dr R. S. Edmondson go my especial thanks, and lastly I must acknowledge Mr B. King, A.I.M.L.T., my helper in all this work.

## References

Ashcraft, K. W. & Leape, L. L. (1970) Candida sepsis complicating parenteral feeding. *J. Am. Med. Ass.* **212**, 454.

Cohen, S. J. & Callaghan, R. B. (1961) A syndrome due to the bacterial colonisation of Spitz-Holtzer valves. *Br. Med. J.* **2**, 677.

Craddock, P. R., Yawata, Y., VanSaten, L., Gilberstadt, S., Silvis, S. & Jacob H. S. (1974) Acquired phagocyte dysfunction—a complication of the hypophosphataemia of parenteral hyperalimentation. *New Engl. J. Med.*, **290**, 1403.

Davidsohn, I. (1962) *In: Clinical Diagnosis by Laboratory Methods.* 13th ed. p. 210 Philadelphia: W. B. Saunders Co.

Dunlop, D. M. & Murdoch, J. McC (1960) The dangers of antibiotic treatment. *Br. Med. Bull.*, **16**, 67.

Freeman, R. & King, B. (1972a) Infective complications of indwelling intravenous catheters and the monitoring of infections by the NBT test. *Lancet*, **1**, 992.

Freeman, R. & King, B. (1972b) NBT test and mycoplasma. *Lancet*, **1**, 962.

Freeman, R. & King, B. (1973) Respiratory tract specimens from intensive care patients. *Anaethesia*, **28**, 527.

Freeman, R., King, B. & Kite, P. (1973a) Serial nitroblue tetrazolium tests in the management of infection. *J. clin. Path.*, **26**, 57.
Freeman, R., King, B. & Hambling, M. H. (1973b) Infective complications of open heart surgery and the monitoring of infections by the NBT test. *Thorax*, **28**, 617.
Freeman, R. & King, B. (1975a) Analysis of catheter tip cultures in open-heart surgery patients. *Thorax*, **30**, 26.
Freeman, R. & King, B. (1975b) NBT tests. *Lancet*, **1**, 104.
Freeman, R. King, B. (1975c) Isolations of aerobic sporing bacilli from the tips of indwelling intravascular catheters. *J. clin. Path.*, **28**, 146.
Freeman, R. & King, B. (1975d) Recognition of infection associated with intravenous catheters. *Br. J. Surg.*, **62**, 146.
Gardner, R. M., Schwartz, Roasanne, Wong, H. C. & Burke, J. P. (1974) Percutaneous indwelling radial-artery catheters for monitoring cardiovascular function; prospective study of the risk of thrombosis and infection. *New Engl. J. Med.*, **290**, 1227.
Gordon, A. M., Rowan, R. M., Chanduri, A. K. F. & Moran, F. (1974) A further application of the nitroblue tetrazolium test. *Br. Med. J.*, **3**, 317.
Leading Article (1971) Nitroblue tetrazolium: a routine test. *Lancet*, **2**, 909.
Levin, J., Poore, T. E., Zauber, N. P. & Oser, R. S. (1970) Detection of endotoxin in the blood of patients with sepsis due to Gram negative bacteria. *New Engl. J. Med.*, **283**, 1313.
Park, B. H., Fikrig, S. M. & Smethwick, E. M. (1968) Infection and nitroblue tetrazolium reduction by neutrophils. *Lancet*, **2**, 532.
Price, D. J. E. & Sleigh, J. D. (1970). Control of infection due to *Klebsiella aerogenes* in a neurosurgical unit by the withdrawal of all antibiotics. *Lancet.* **2**, 1213.
Rodman, N. F., Wolf, R. H. & Mason, R. G. (1974) Venous thrombosis on prosthetic surfaces: evolution and blood coagulation studies in a non-human primate model. *Am. J. Path.*, **75**, 229.
Rosenthal, A. (1958) Follow up study of fatal penicillin reactions. *J. Am. Med. Ass.*, **167**, 1118.
Segal, A. W. (1974) Nitroblue tetrazolium tests. *Lancet.* 2, 1248.
Sallam, I. A., Mackey, W. A. & Bain, W. H. (1970) Prophylactic antibiotics in intensive therapy: experience in a cardiac surgical unit. *Br. J. Surg.*, **57**, 722.
Weinstein, L. & Rubin, R. H. (1973) Infective endocarditis—1973. *Prog. cardiovasc. Dis.* **16**, 239.

# 2 | Assay of Antimicrobial Agents

## D. S. REEVES and M. J. BYWATER

Ever since the introduction of antimicrobial drugs there has been a need for measuring their concentration. Nowhere has this been more pressing than in their manufacture and subsequent quality control as distributed phamaceutical products. As a consequence a vast wealth of expertise and literature has been built up but much of it is couched in a jargon which will discourage its assimilation by the average worker in a routine clinical microbiology laboratory. Furthermore, many of the techniques described and discussed in considerable detail in the industrial literature are designed for the very high precision assay of large numbers of samples, often over a small range of concentrations and usually presented as high concentrations in aqueous media free of interfering substances. It is not surprising then that methods of assay more suited to samples from patients have been dispersed among the literature and to our knowledge more than a brief review of the technicalities of these assays has not appeared before. Those requiring a shorter account are referred to Blair et al. (1974), and the excellent section on methods by Waterworth in Garrod et al. (1974).

In the present review the authors have attempted within the confines of space to give a comprehensive account of the assays available for estimating the concentrations of antimicrobial drugs in samples from patients. For exact technical details the reader is referred to the literature cited. In the case of agar plate diffusion assays so little practical advice is published on techniques suited to the clinical laboratory the authors make no apology for the detail given.

Before starting on the discussion proper of assay techniques, the behaviour of antimicrobial drugs in the body will be reviewed briefly together with its relevance to the need for assays.

## Pharmacology and The Need for Assays

To reach their site of action, which is often in the tissues, antimicrobial drugs must run the gauntlet of a number of barriers, just how many depending on the route of administration. In addition they will be treated as foreign substances to be eliminated from the body and will be subject to the processes of metabolism and excretion (Fig. 2.1).

To be delivered to the site of activity a drug must enter the blood. Exceptions to this are locally administered active agents. Sometimes the drug will be given directly into the vascular compartment by bolus intravenous injection or by infusion. More usually the route of administration is extravascular, either by injection or oral administration.

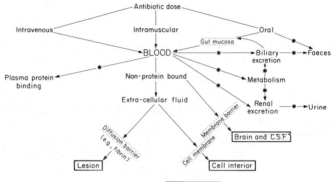

Key: (i)   possible site of activity, e.g. ⎯LESION⎯
     (ii)  ⎯⎯⎯→ —route of access
     (iii) ⎯ * → —route of removal from access

**Fig. 2.1.** *Schematic representation of factors which influence the access of antibiotics to the site of infection in the tissues.*

### Oral Administration and Intestinal Absorption

The vast majority of courses of antibiotics prescribed in the western world are for oral administration. The reasons are comfort and the convenience of self medication. It must be remembered however that prescription is not proof of ingestion and that patients often forget to take their drugs at the times prescribed or even at all. After oral ingestion the absorptive process of many drugs is not reliable. The penicillin group in particular offend in this respect and it is for this reason that the serum concentrations of penicillin should always be ascertained after a change from parenteral to oral therapy during the management of subacute bacterial endocarditis. Similarly, the absorption of ampicillin is too small to obtain reliable serum levels high enough to treat tissue infections with sensitive coliform bacteria, and intramuscular therapy is essential. The poor absorption of the *broad-spectrum penicillins* ampicillin and carbenicillin have been improved by esterification. After absorption the compounds are de-esterified by naturally occurring esterases with release of the free parent drug, the ester moiety being metabolized and excreted. Examples are carfecillin, the phenyl ester of carbenicillin, and talampicillin, the phthalidyl ester of ampicillin. Free carbenicillin and ampicillin, respectively, are released into the body after absorption. Oral *cephalosporins* are well absorbed. Many of the *tetracycline* group are in particular irregularly absorbed from the gastro-intestinal tract and serum levels should be checked during therapy for serious infections.

*Trimethoprim*, most *sulphonamides* and *chloramphenicol* are all more lipid soluble than the highly polar penicillins and cephalosporins and are well absorbed. Quite small changes in molecular structure may dramatically modify pharmacological properties. The substitution of a chlorine atom for a hydroxyl group to produce *clindamycin* from *lincomycin* improves its absorption, alters its distribution in the body, and increases its antimicrobial activity.

Some antibiotics, for example those of the *aminoglycoside* group, are virtually non-absorbed from the gut. Having said this, in patients with poor renal function and perhaps abnormal intestines, sufficient antibiotic can be absorbed to accumulate and cause a toxic effect. Regular serum assays will reveal this phenomenon should it occur.

In recent years it has become apparent that not all preparations of the same antibiotic are necessarily equally well absorbed and so unexpected variations in serum levels occur. The extent to which these differences in bioavailability occur is still under investigation and is probably less important with antibiotics than the more dose-critical drugs as digoxin or anti-diabetic preparations.

Finally, mention should be made of patients with gastrointestinal disease (for example, coeliac disease or partial gastrectomy) in whom the normal absorption of orally administered drugs cannot be assumed.

## Availability from Parenteral Administration

As mentioned above, to be distributed to its site of action a drug must enter the blood. Absorption from extra-vascular (intramuscular and subcutaneous) injection sites is usually reliable. However, conditions causing poor tissue perfusion such as arterial hypotension or local sclerosis due to repeated injections may delay absorption. If drug elimination remains unimpaired low serum levels will result. This risk can be obviated in the acutely ill patient by checking that serum levels are satisfactory, or by direct intravascular administration.

Intravenous administration by bolus injection and by slow infusion are not synonymous, particularly when a drug rapidly cleared by the kidneys is being given. When renal function is normal it is possible to conceive that a slow continuous infusion of a dose would fail to produce adequate levels, due to the rate of elimination matching the rate of administration, whilst the same dose given over a short period would produce a high blood level which, although declining, would persist for some time. Even when renal function is impaired it would take some time to reach optimum blood levels using a slow infusion. From the purely microbiological view-point there is little or no evidence to suggest that continuous inhibitory levels are any more effective in curing infection than intermittent peak levels. All the evidence suggests that intermittent peaks of drug concentration produced by intramuscular administration are effective as therapy although how frequent these peaks need to be is conjectural.

## Distribution

With the exception of urine and perhaps cerebrospinal fluid, the last point on the journey to its site of action at which the concentration of an antibiotic is conveniently measured is the blood. Drug not bound to plasma protein or blood cells can penetrate freely into the extracellular fluid (ECF) provided it does not exceed a certain molecular size, and most antimicrobial agents are relatively small molecules. The major limiting factor for transport into the ECF is protein binding. In the plasma free and bound drug are in equilibrium, the proportions depending on the degree of protein binding. Only the free drug is able to diffuse into the ECF and thus protein binding is an important determinant of concentration in the ECF (Fig. 2.2). ECF contains very little protein and therefore drug in it will be only slightly bound, if at all. At first glance this may give a gloomy picture on the activity of the antibiotic outside the blood compartment. It should be realised however that only the free drug in the plasma is able to exert an antibacterial effect in that site and therefore the activities of the antibiotic in the ECF and plasma would be similar, assuming that no other factors affected them. The blood brain barrier is an important exception to the ready diffusibility of drugs from the blood to the extravascular compartment, the capillary endothelial junctions being much tighter. As a consequence only lipid soluble drugs readily enter the brain.

From the extracellular fluid the next barrier to entry into the cells of the tissues themselves is the cell membrane. The cell membrane consists of layers of protein and lipid molecules. Passage of drugs is

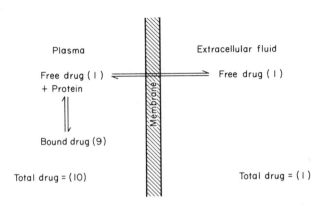

**Fig. 2.2.** *Theoretical effect of serum protein binding on the distribution of a drug between the blood plasma and the extra-cellular fluid.*

usually by passive diffusion down a concentration gradient, active or facilitated transports being unusual for exogenous substances. The rapidity with which equilibrium of concentrations takes place is determined by the diffusibility of the drug, which is in turn determined largely by its lipid solubility. Molecules with polar groups (–OH, –COOH, –NH$_2$, etc) are more hydrophilic and less lipid soluble. Lipophilicity is increased by non-polar groups (CH$_3$–, CH$_3$CH$_2$–, phenyl, etc). Polarity is increased by ionization which is in turn determined by the pKa of a group and the ambient pH. Very small ionized molecules (e.g., urea) are able to penetrate the membrane through the narrow aqueous channels which are thought to exist. The molecules of most antibacterial agents are ionizable and many are highly charged at the typical pH values found in the body (e.g., penicillins and cephalosporins). Some (e.g., sulphonamides, trimethoprim, lincomycin) are partially ionized at plasma pH and become more or less so when secreted into fluids of a different pH (e.g. saliva, bile, prostatic fluid) which will in turn alter markedly the ratio of concentrations across the membrane (Fig. 2.3).

Diffusion of a solute is hastened in facilitated transport along a concentration gradient apparently without the expenditure of energy although the concentrations will be equal at equilibrium. There has been no clear demonstration of facilitated transport being involved in drug distribution. Active transport will move solutes against a concentration gradient with the expenditure of energy. There is no evidence that antibiotics are distributed in this way but drugs resembling normal dietary constituents (e.g. α-methyldopa) may be actively absorbed from the bowel.

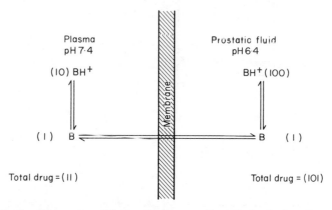

**Fig. 2.3.** *Conditions at equilibrium for the secretion of a basic drug from plasma into an environment of a lower pH, for example prostatic fluid. The drug is assumed to have a pKa of 8·4 (50 per cent ionized at pH 8·4) and therefore will be present in an ionized/unionized ratio of 10 at pH 7·4, and 100 at pH 6·4. The acid environment is assumed to be prostatic fluid with, for convenience, a pH of 6·4. Note that at equilibrium only unionized drug is present in equal concentrations across the membranes. An acidic drug would have a lower concentration in the prostatic fluid than in the plasma because it will be less ionized at the lower pH.*

A useful concept in drug distribution is the *apparent volume of distribution*. This is the volume in which the total amount of drug in the body would be uniformly distributed to give the observed plasma concentration. For a total amount of administered drug $(Q)$ the apparent volume of distribution $(V_d)$ for a plasma concentration $(c)$ would be given by: $V_d = Q/c$.

A very diffusible drug would have a large volume of distribution while a drug confined to the vascular compartment would have a small one. In this context it is important to remember that the plasma, extracellular and cellular waters constitute some 4 per cent, 20 per cent and 60 per cent of the body water, respectively, so a small amount of penetration into the cells would have a large effect on $V_d$. The volume of distribution is only a ratio and has no anatomical meaning. None the less it is a useful starting point for pharmacokinetic manipulations. For further reading the account to be found of drug distribution in LaDu et al. (1971) is to be recommended.

## Elimination

Excretion in the urine is the major final pathway of elimination of most drugs. Some antibacterial drugs are excreted almost entirely unchanged (e.g., aminoglycosides) when over 90 per cent of a parenterally administered dose can be recovered as microbiologically active drug in the urine. Such drugs are filtered into the tubular fluid and pass on into the urine with little or no tubular reabsorption. Lipid soluble antibacterials on the other hand are often extensively reabsorbed from the renal tubules, and transformation by liver enzymes into more polar metabolites is essential if the drug is not to persist indefinitely in the body. The process of metabolism is often carried out in two steps. In the first, polar groupings are introduced (e.g., oxidation, hydroxylation, dealkylation). Conjugation then takes place with lipophobic moieties to form, for example glucuronides, acetylates, sulphates. Conjugation may occur without previous biotransformation, and both processes may occur simultaneously. Examples of commonly used antimicrobial agents which undergo extensive biotransformations are sulphonamides, trimethoprim, nitrofurantoin, chloramphenicol and some tetracyclines.

The ability to biotransform drugs is genetically determined and with most drugs the natural spectrum of this ability produces a variation of blood levels when a fixed dose is given to a population of individuals. These differences in plasma levels may be important for dose-critical drugs such as those acting on the central nervous system. Occasionally a genetic abnormality produces a dramatic effect such as the lack of plasma pseudocholinesterase giving scoline apnoea.

Previous exposure to drugs can induce the hepatic microsomal enzymes responsible for drug metabolism. It is fortunate that the large majority of antibacterial drugs are neither significantly affected in this way or are themselves responsible for such an induction. Induction of liver enzymes are responsible for a number of well-known drug interactions such as that between barbiturates and anticoagulants.

Renal excretion and biotransformation can be competing agencies for the removal of active drug. When renal function is diminished excretion is delayed and more extensive metabolism occurs leading to a low recovery of active drug in the urine. This can assume a practical importance since the concentration of microbiologically active drug in the urine can fall to levels at which a urine infection might not be eradicated. Examples are nitrofurantoin and chloramphenicol. In patients with severely impaired renal function, metabolism may assume importance with drugs in which it normally constitutes a negligible route of elimination.

Antibiotics predominantly excreted unchanged by the kidneys will accumulate in the body if given in unmodified dosage in the presence of renal failure. In the case of some penicillins such as carbenicillin which need normally to be given in high dosage and are virtually non-toxic this accumulation can be a positive advantage since it allows a lower dose to be given. Where, however, the antibiotic has toxicity related to blood levels (e.g., ototoxicity of aminoglycosides) dosage must be modified, and blood level monitoring can help in finding the correct regime for a particular patient.

The excretion of antibacterial drugs which are only partially ionized at the prevailing pH of tubular urine can be markedly affected by changes in the pH of the urine within physiological limits. Conversely, drugs which are highly ionized already (penicillin and cephalosporin) cannot be influenced. For example, the excretion of the ultra–long–acting sulphonamide 'sulfametapyrazine' can be hastened by urinary alkalization (Fig. 2.4). The increased quantity of sulphonamide excreted, however, only represents a small fraction of the total amount of drug in the body and the shortening of its biological half-life is not dramatic.

## Relationship of Blood Levels to Efficacy

It is rare to have infecting micro-organisms primarily in the blood and it is appropriate therefore to examine how blood concentrations of any drug relate to the elimination of organisms outside the vascular compartment. In urinary infections the concentration of most antibacterials will easily exceed that required to inhibit the organism. The reverse is true of most other sites where the concentrations are expected to be lower than those found in plasma. Serum concentrations as measured by most assay methods either tend to measure or actually measure total drug (free and protein bound). Even drugs which are highly protein bound (e.g., 95 per cent bound, 5 per cent free) may assay as if less bound (e.g., 75 per cent bound, 25 per cent free) (*see* the section on choice of medium for working standards). With all antibacterials it is perhaps wisest to assume that the assay has measured total drug and estimate what is free by reference to published values for protein binding.

Many infective lesions are outside the tissue cells and in the early stages at least might be anticipated to be bathed in ECF in which the

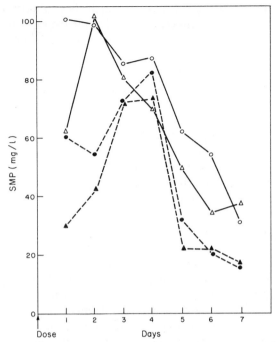

**Fig. 2.4.** *Effect of alkalinization of urine on the excretion of sulfametapyrazine (SMP) and its acetyl derivative in two volunteers (● and ▲). Alkali (sodium bicarbonate) given days 3 and 4. Note the marked effect on the more lipid soluble parent drug caused by the increasing ionization at high pH values decreasing tubular reabsorbtion. ○——○ and △——△, Aceylated sulfametapyrazine; ●–·–·–● and ▲————▲ Free sulfametapyrazine. (Reeves, 1975).*

concentration of antibiotics would be crucial. Later, pathological changes take place (exudation of high protein content fluid and perhaps fibrin deposition) which might have a profound effect on the local distribution of a drug. To mimic these changes in an assay system with any degree of reliability would be impossible but they must be taken into consideration when deciding on what is a suitable blood level since they can act as a barrier to drug diffusion.

Another factor which can play a part in the cure of an infection is bactericidal activity. Normally bacteristatic action is sufficient to assist the defence mechanisms of the body in removing bacteria from the tissues. However, there may be local factors (e.g., a fibrin barrier to phagocytes, or a foreign body) or general deficiencies in immune mechanisms which make bactericidal activity essential. Most antibiotics which possess bactericidal activity are able to kill bacteria only at concentrations higher than those at which they inhibit. If the minimum inhibitory (bacteristatic) concentration for the offending microbe is measured or deduced a further allowance for bactericidal activity will need to be made.

It is for the above reasons, together with an allowance for any inaccuracies of measurement, that the usual recommendation for blood levels of penicillin for the treatment of subacute bacterial endocarditis is about eight-fold higher than the minimal inhibitory concentration. Such a generous allowance over the inhibitory concentration cannot be made however in the case of potentially toxic antibiotics. In spite of its clinical use for ten years it is surprising therefore that only recently have definite recommendations been made with regard to the minimum peak blood levels for gentamicin for effective treatment (see Reeves, 1974).

Some infections are predominantly intracellular and diffusibility of drug into the cells becomes of paramount importance. It is not possible to make direct measurements under routine circumstances but the likely behaviour of any agent can be deduced from a knowledge of its physicochemical properties and pharmacokinetic behaviour. A well known example is the effectiveness of chloramphenicol, a non-ionized and highly diffusible molecule but with bacteristatic properties, in typhoid fever as compared with the bactericidal agent ampicillin. Other antibiotics which are as effective in vitro against the causative bacterium (e.g., tetracyclines) fail to work in vivo. Sulphonamide/trimethoprim mixture is one of the best treatments for brucellosis, another intracellular infection, when again many agents active in vitro are ineffective.

There are many other sites of infection where the relationship between blood and local concentrations require special consideration. Examples are cerebrospinal fluid, neural tissue, bone, bronchial and other respiratory secretions, bile, the fetal circulation, and prostatic fluid.

## Relationship of Blood Concentrations to Toxicity

Although the commonest reason for a clinical laboratory to assay blood for an antibacterial drug is to ensure that sufficient is present, another important use of assays is to check that toxic concentrations are not being reached. With some toxic effects (e.g., allergic reactions to penicillins) there is no relationship to blood concentration apart from the mere presence of the drug. The penicillins and cephalosporins are remarkably non-toxic but levels of over 100 mg/l should be exceeded with caution as penicillin encephalopathy and cephaloridine nephrotoxicity may occur. There is a much smaller margin between toxicity and efficacy with the aminoglycoside antibiotics. In spite of there being 'recommended' peak levels above which toxicity is thought to be more frequent, there is little evidence to define either what these levels are or whether they even exist. On an 8-hourly dosage regime of genta-mycin designed to produce adequate peak levels, it is currently thought that the technically accurate trough should be 2 mg/l or less. Higher levels would suggest depression of renal function known to be associ-ated with aminoglycoside to toxicity.

## Assay Method for Antimicrobial Drugs

Assays of antimicrobial agents can broadly be divided into three groups based on the increasing order of accuracy required:
  (1). Assays for the control of chemotherapy.
  (2). Assays for the pharmacological assessment of antimicrobials.
  (3). Assays for quality control in manufacturing processes.

Clinical laboratories are almost entirely involved with the first group, although occasional needs for the second group will be encountered by some. This chapter is intended largely for clinical laboratory use and the procedures described and discussed will therefore be almost entirely confined to assays of the lowest permissible order of accuracy.

Although a more general discourse is implied by the title, our dis-cussion of assay methods will be almost entirely confined to those of antibacterial agents. This specialization is justified by the requirements of clinical laboratories within the confines of this chapter. Antiviral agents are still in a developmental stage. Assays of antifungal agents will be mentioned briefly. Apart from assays for streptomycin, assays of anti-tuberculous agents are rarely requested of routine laboratories. There will also be an emphasis on assays for aminoglycoside antibiotics. particularly gentamicin, as they constitute by far the greatest number of individual routine requests for assay and are also the most critical with regard to accuracy. Again our justification for this unbalance is expediency.

Before considering assays suitable for use in clinical microbiology laboratories it is first necessary to outline the requirements of the various factors which make an assay suit this purpose.

### *Outline of Requirements of an Assay*

*Accuracy.* The accuracy of any particular assay may be judged on how near the result (i.e., the concentration of antimicrobial agent observed) is to the true concentration of antimicrobial agent in the sample and with what degree of reliability (i.e., the reproducibility of the assay). The requirements of accuracy vary according to the drug to be assayed and the clinical circumstances under which the samples were taken. For example, in ascertaining that sufficient penicillin is in the blood during the treatment of bacterial endocarditis an assay which reliably put the observed concentration between 2 and 4 mg/l would suffice. On the other hand assays of gentamicin, an antibiotic which has relatively narrow margins between efficacy and safety, should distinguish reliably between concentrations of 8 and 12 mg/l for peak blood levels, and 1·5 and 2·5 mg/l for pre-dose blood levels.

Thus, in the context of controlling aminoglycoside therapy, assays should be reliably accurate to at least ±30 per cent, and to give a margin for error an accuracy of better than ±25 per cent would be satisfactory. Naturally, if an assay was usually ±15–25 per cent accurate, an occasional result may be less accurate than ±25 per cent. To allow for this accuracies may be expressed as 95 per cent confidence limits. Thus an accuracy of ±25 per cent at the 95 per cent confidence limits would mean that about 19 out of 20 assays could be expected to fall within the ±25 per cent limits, and most assays will be more accurate than ±10 per cent, based on the normal distribution. The remaining 1 out of 20 assays would be less accurate than ±25 per cent. Assays for pharmacological purposes would need to be more accurate since important general conclusions regarding a drug may be drawn from a relatively small number of observations.

The accuracy needed for any particular assay can only be decided after due consideration of its intended use. Although it might be thought that all assays should be highly accurate this achievement would be difficult and wasteful of resources. In the clinical laboratory the diversion of these resources from other valuable work to produce unnecessarily accurate assays could thus act to the detriment of patients.

*Specificity.* An assay may be affected by the presence of other substances so as to produce an inaccurate result. This usually arises from the presence of other antimicrobial drugs which interfere in a microbiological assay. Interference may also be produced by breakdown or metabolic products of a single agent. For example, carbenicillin cannot be assayed in a microbiological system using a Gram-positive indicator organism because it spontaneously breaks down to benzyl penicillin which exerts a greater anti-Gram-positive action than the parent drug. With chemical assays an inactive metabolic product may be measured, as in one colorimetric assay of chloramphenicol. For clinical purposes it would be important to check carefully that any antibiotic measured by a non-microbiological procedure was in fact microbiologically active. There is also something to be said for using an indicator organism which resembles, at least to some extent, the infecting organism for which the patient is being treated, since in the event of non-specificity the assay result would at least be relevant to the adequacy of therapy.

*Speed.* Assays for controlling therapy often have a requirement for producing answers quickly. In the case of a drug which is non-toxic (e.g., benzyl penicillin) a large excess can be given as initial therapy so as to ensure the achievement of adequate blood levels. Thus speed of producing a figure for the peak blood level is not essential and an assay taking overnight incubation is satisfactory. When however, the antibiotic is an aminoglycoside which must not be given in overdosage then an assay to ensure adequate peak blood levels should with advantage produce an answer before the next dose is due, which is often about 7 h after taking the blood.

A very quick assay (less than 2 h) also has an advantage in that, except with an overnight assay, the result is more likely to be forthcoming during the working day. In our experience a 5-h assay often needs to be read during the evening.

*Sample size.* Repeated blood samples are sometimes necessary for the control of chemotherapy. While the total volume of blood taken may not be important in adult medicine it is clearly so for children and especially neonates. During pharmacological studies many samples will often be taken from each subject and again volume is important. For a microbiological assay 0·3 ml of serum is usually a minimum because of the replicates needed. Some chemical assays need larger volumes because of lack of sensitivity (e.g., the spectrofluorimetric assay of trimethoprim) while others need only 100 µl or less (e.g., radio-immunoassay of gentamicin).

*Sensitivity.* Sensitivity is the capability of an assay to measure the lowest relevant concentration of antimicrobial agent to be found in a sample and in most instances this means blood samples. Sensitivity is usually adequate with microbiological assays but can be a problem with nonmicrobiological assays.

*Cost.* In the situation of competing demands on resources cost is an important factor. As in the allocation of all health-care resources a responsible decision on the benefit to patients related to cost may be required if the assay method is expensive. Fortunately most of the methods considered here are inexpensive in the context of western medicine.

## Assay Methods Available

An enormous variety of techniques have been described for assaying antimicrobial drugs but many of them are not applicable to clinical estimation since they are seriously deficient in at least one of the requirements discussed above. Those suitable for use in a clinical laboratory may be divided into *microbiological assays*, defined as assays where the indicator of drug concentration is an effect on the growth of a population of organisms, and *non-microbiological assays*.

### Microbiological Assays

*Types of microbiological assays*
(1). Agar diffusion assays
    —plate diffusion
    —vertical diffusion.
(2). Broth or agar dilution assays.
(3). Turbidimetric assays.
(4). Urease assay.

Microbiological assays are usually characterized by adequate sensitivity and relevant specificity in that they can be designed to measure the very activity sought for in the in vivo situation. Being biological methods they are all subject to biological variation and therefore require internal replication to minimize error from this source. With adequate replication and good technique they are generally capable of achieving accuracies of $\pm 15$ per cent–$\pm 25$ per cent without being excessively burdensome to a clinical laboratory. Higher accuracies would require greater numbers of replicates and more precise technique. The time within which a result is available is limited in diffusion assays by the rate of growth of the indicator organism and by the rate of diffusion of the antimicrobial drug through agar. Typical incubation times for rapid assays are 4–6 h. Conventional diffusion assays are usually incubated for 18 h.

*Agar diffusion assays.* These may be plate assays, or vertical diffusion assays. *Agar diffusion assays in plates* have been brought to a high state

of development because of their applicability to controlling the manufacture of both antibiotic and non-antibiotic substances (Kavanagh,
1963, 1972a). When necessary assays of very high accuracy can be
achieved, but their appeal to the clinical laboratory is in the simplicity
of the method, the relative cheapness of materials and equipment, and
the ability to assay a variety of agents with only small changes in methodology. Because we believe that many laboratories subscribe to this view
some considerable detail on agar plate diffusion will be presented
below. Having appraised the value of these assays enthusiastically it
must be said that with poor technique extremely inaccurate results are
almost inevitable, a fact borne out by quality control surveys (Reeves
& Bywater, 1975).

*Vertical diffusion assays.* First described in a classical paper by
Mitchison and Spicer (1949) for the assay of streptomycin in body
fluids. These assays are performed in narrow-bore parallel-sided tubes
in which the sample (serum or other body fluid) is layered on to a
column of a nutrient agar containing the indicator organism. After
incubation a zone of inhibition develops in the agar adjacent to the
agar/sample interface and is measured with a vernier microscope. The
concentration of drug in the sample is read off a standard curve prepared from a set of tubes containing standard concentration in serum
or other body fluid. The method is adaptable to all aminoglycosides
but does not appear to be widely used. Because of the precise manner
in which the zones are measured vertical diffusion assays might perhaps
be more accurate than plate diffusion assays, in which a major weak
point is zone reading, as practised by many laboratories.

*Dilution assays.* In these assays the substance to be measured is
diluted in a nutrient medium and an end-point detected by the growth
or otherwise of an indicator organism. The concentration is found by
comparison to a similarly prepared row of dilutions with known antibiotic concentrations. The usual practice is to employ 2-fold dilutions.
but intermediate dilutions to form an arithmetic or semi-arithmetic
series can also be used although the answers are always in the form of
a discontinuous variable. Using a 2-fold series the steps are too wide
apart to give accuracies of better than $\pm 50$ per cent. With closer dilutions the end-point may not be qualitatively sharp thus requiring the
extra labour of some quantitative test, for example, involving viable
counting. Doubling dilution assays are adequate for some purposes
in controlling therapy when accuracy is not essential. They also form
the basis of the useful test of measuring the inhibitory and bactericidal
activity of a patient's serum against his own infecting organism.

*Turbidimetric assays.* These are assays in which the antimicrobial substance inhibits the growth of an indicator organism in both. A series of standard concentrations can be prepared, each producing a different turbidity measured at the end of a period of incubation. Thus a graph of final turbidity versus concentration can be prepared and the concentration of the unknown determined from it by its own final turbidity. The method has the advantage that meaningful differences in turbidity can be detected after only 2 h incubation with the appropriate indicator organism (Fig. 2.5). Although turbidimetric assays are widely used for industrial quality control when aqueous solutions are being used (*see*, for example, Kavanagh, 1972b), they have been unpredictable in accuracy when used with complex samples such as serum, presumably due to unidentified inhibitors (usually) or stimulants of bacterial growth (Noone, 1973). Furthermore, because of the discrete nature of each standard and test an unforeseen technical variable (e.g., an inadequately cleaned tube, or a slight variation in inoculum size) may produce an individual wayward result and thus replication is essential.

Finally, the range of concentration over which there is partial inhibition of growth as opposed to full growth or no growth is small (typically 2–10 fold) and this means that clinical samples would need to be assayed in dilutions to ensure a result on the first occasion. Because of all these problems turbidimetric assays have not found favour for clinical laboratory use.

Similar objections could also apply to other growth inhibition assays in fluid culture where the rapid detection of differences in bacterial growth by bioluminescence, microcalorimetry, or electrical resistance have been suggested.

*Urease assay.* This method was introduced and developed by Noone and his colleagues (1971). The assay technique is based on the induction of a urease enzyme in *Proteus mirabilis*, when incubated in a urea broth, and the subsequent alkalinization of the medium by the production of $NH_4^+$ following urea hydrolysis. The pH change is measured to an accuracy of 0·01 units. The organism, previously unexposed to urea, is incubated in the urea broth together with the serum sample or gentamicin standards. Any aminoglycoside present in the serum will slow the rate of protein synthesis and thus reduce the pH change. By including a serious of gentamicin standard dilutions and also a series of standards plus the serum sample it is possible to overcome any effects due to differing buffer capacity of the serum samples. The minimum quantity of serum required is about 1·5 ml for the standard technique and 0·4 ml for a modified technique. The method would appear to be

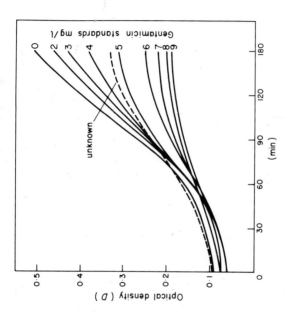

**Fig. 2.5.** *Rapid assay of gentamicin by turbidimetry. The concentrations given are the concentrations of gentamicin in the standard sera. Each curve represents the change in optical density (D) of a culture of Proteus mirabilis shaken in a cuvette at 37°C. By inference, the concentration of gentamicin in the unknown sample was between 4 and 5 mg/l. (Breeds & Reeves, unpublished data).*

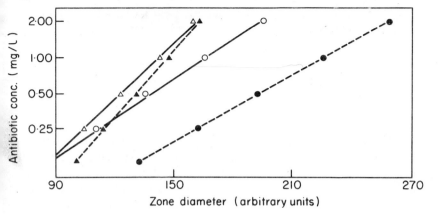

**Fig. 2.5.** *Rapid assay of gentamicin by turbidimetry. The concentrations given are the concentrations of gentamicin in the standard sera. Each curve represents the change in optical density (D) of a culture of* Proteus mirabilis *shaken in a cuvette at 37°C. By inference, the concentration of gentamicin in the unknown sample was between 4 and 5 mg/l. (Breeds & Reeves, unpublished data).*

less accurate in the critical range of pre-dose concentration of gentamicin of 1·5–3·0 mg/l than at higher concentration. A summary of the method is given by Noone et al. (1975).

The urease assay is apparently unaffected by the presence of penicillins, cephalosporins, co-trimoxazole and rifamycins, and a result is obtained in 2 h. There is no doubt that in experienced hands the assay can produce accurate results (± 15 per cent). As with all microbiological assays less proficient laboratories have not always produced good results, and the method does not generally offer a significant improvement in accuracy when compared to agar plate diffusion methods as used in routine clinical laboratories (Reeves & Bywater, 1975). A disadvantage of this more specialized type of assay is that it is not applicable to all the likely estimations of antimicrobial drugs requested of a laboratory. Thus expertise in more than one type of technique is required.

## Methodology for Plate Diffusion Assays

Plate diffusion assays have a broad applicability in the clinical laboratory and are, in fact, the most frequently used method for gentamicin assays (Reeves & Bywater, 1975). It seems appropriate therefore to discuss the technique of performing them in some detail, identifying critical points which, in the authors' experience, are major sources of error.

## Preparation of Stock Standards

A stock standard is a solution of accurately known concentration from which the working standards used in the actual assay are prepared. Errors in the preparation of standards are frequent in clinical laboratories. They may arise initially from the choice of inappropriate standard material. Standard material must be in a microbiologically active state—a well known trap is the use of chloramphenicol for injection (an inactive ester) instead of the active but poorly water soluble pure substance. The potency (i.e., the activity of the compound per measured weight) must be accurately known, and reference to the manufacturer will usually help in this. Some antibiotics, for example benzyl penicillin, have virtually a 100 per cent potency while others, for example gentamicin, may have a potency as low as 60 per cent i.e., 600 mg per 1000 mg of weight. Confusion can arise because the substance is provided as a salt but the potency is expressed as the free base or acid. Allowance for the salt radical must be made during weighing. The contents of an ampoule prepared for injection must never be assumed as the indicated weight, volume, or solution strength. Dry powders in ampoules usually have an overage of weight, and solutions may have overages of volume and concentration. The exact concentration of a solution (as opposed to the labelled concentration) can be found by referring the batch number to the manufacturer. A preparation for oral dosing should never be used since it may not be an active compound and may also contain a substantial proportion of inert filling material.

A further complication regarding potency is hygroscopicity. For example, both gentamicin and tobramycin powders can be dried at 120°C (240°F) and cooled in a dessicator. On weighing at normal ambient humidities they rapidly acquire water and the alteration in weight can easily be followed on an accurate balance. The more rapidly a weighing is made the less will be the error from this source. Alternatively, weighings can be made at timed intervals and the original truly dry weight deduced by backwards extrapolation. A high quality balance with adequate damping is essential to ensure a rapid and accurate weighing. It should be capable of weighing to an accuracy of better than 0·1 mg, and at least 50 mg of substance should be used.

After weighing the substance it is completely transferred to a volumetric flask of appropriate volume by washing and made up to the mark. Water or preferably a buffer is used for solution. The pH and molarity of buffer chosen will be governed by such considerations as the stability of the antimicrobial substance at different pH values and the final pH of the working standards. Some substances will keep well, particularly if deep-frozen or stored in liquid nitrogen, while others are more labile, and a fresh stock standard must be prepared on the day of the assay. If stored, the standard should be labelled with its concentration, batch number of substance and date of preparation, as well as its identity.

## Preparation of Working Standards

Working standards must be prepared in a medium appropriate to the body fluid to be assayed, particularly with regard to pH and protein

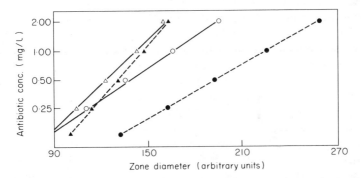

**Fig. 2.6.** *Microbiological assay of cloxacillin and benzyl penicillin arranged to show the effect of protein binding.* ▲---▲: *benzyl penicillin in buffer.* △—△: *benzyl penicillin in serum.* ●--●: *cloxacillin in buffer.* ○—○: *cloxacillin in serum.*

content. With some antimicrobial drugs (i.e., gentamicin) even quite small differences (pH 0·1) between the pH of the standards and the sample can lead to marked loss of accuracy (Elfving & Pettay, 1974). The difference made by the use of protein-containing standards again depends on the particular antibiotic being assayed. Highly protein-bound drugs (e.g., fusidic acid 97 per cent bound, cloxicillin 95 per cent) will be more affected than those with relatively low protein binding (e.g., benzyl penicillin 60 per cent, ampicillin 18 per cent). (Fig. 2.6).

Standards for *serum assays* should be made in serum, preferably antibiotic-free pooled human serum drawn from donors tested for

C

absence of Australia antigen. When using serum from less defined sources it should be checked in some sensitive microbiological system for absence of antimicrobial activity. No serum which is not fresh can be assumed to have a pH of even near to 7·4. In the absence of a suitable source of human serum, animal serum may be used provided it gives demonstrably satisfactory results in the assay system being employed.

Standards for *urine assays* are prepared in a buffer of a suitable pH. Urine samples are diluted in the same buffer, which must be of a strength great enough to ensure that all the sample dilutions have the same final pH.

Standards for other body fluids can usually be prepared in buffer as their protein content is often low, particularly in the main binding substance, albumen. Examples are cerebrospinal fluid and bile. In any event, the recovery of assayed drug can be established by adding a known amount to antibiotic-free samples of the fluid to be assayed.

The *number of working standards* is an important consideration. In industrial assays where the expected range of concentrations is very small (less than $\pm 15$ per cent) two standards are adequate. In clinical assays the range of sample concentrations is much greater and it is often convenient to be able to cover the expected range without dilution of samples. More standards are therefore needed to cover the range to be assayed. Extrapolation above or below their range is not permissible. A good plate assay will often be near to linear over a range of 5 dilutions placed at 2-fold intervals. Attempts to extend the range further may result in a very non-linear assay and dilution of samples would be a more satisfactory method. In many of our own assay methods the lowest standard is one which contains the smallest concentration to produce a clear, well-readable zone of inhibition. For convenience of handling the zone-size data virtually all our assays have a further 4 standards going upwards by 2-fold increases in concentration. To have fewer standards covering the same range of concentrations might well result in less accuracy since a random error in the zone size, say, of one of 3 standards would often be indetectable (*see* Determination of unknown concentration).

The working standards are prepared from the stock standard by careful volumetric dilution using high quality glassware. The top working standard is prepared first in the final medium, taking care that the volume of aqueous stock standard added does not result in a significant dilution of the medium itself. The remaining standards are prepared by dilution from the top standard and perhaps an intermediate dilution. A single 'rogue' working standard might then be easily apparent by persistently failing to align on the standard curve.

Serial doubling dilution should not be used since a consistent error at each dilution might result in a smooth and thus indetectably inaccurate standard curve. Ideally small volumetric flasks and pipettes which run out between marks should be used for these final dilutions. As an illustration of a typical preparation of standards the following procedure might be used for a gentamicin assay: gentamicin sulphate powder (314·8 mg) previously dried and cooled in a dessicator is quickly weighed and transferred completely to a 200 ml volumetric flask (Grade A). The potency of the powder was 635 mg gentamicin base to 1000 mg weight and the stock solution therefore had a concentration of 1000 mg/l expressed as gentamicin base when made up to the mark. The stock solution was then diluted to the top working standard of 20 mg/l by measuring 1 ml into a 50 ml volumetric flask with a volumetric pipette and made up to the mark with pooled human serum. The 10 mg/l and 5 mg/l standards were made by dilution using 25 ml and 5 ml volumetric pipettes respectively and 50 ml and 20 ml volumetric flasks. The 2·5 mg/l and 1·25 mg/l standards were similarly prepared from the 5 mg/l standard.

## Condition and Transport of Samples

Most samples to be assayed will be sera from clotted blood. Samples in other forms should not be assumed to be synonymous with serum as they may give different results when assayed against standards in serum. For example, aminoglycoside assay results are much affected by heparin or citrate. Body fluids other than serum may have low protein contents, for example, cerebrospinal fluid, bile or urine, and should not necessarily be assayed against serum. If the concentration of drug permits they should be diluted in buffer and assayed against buffer standards. Fusidic acid assays in particular need careful dilution and reference should be made to the manufacturer's literature.

Transport and other delays can lead to a loss of potency of labile antibiotics, and serum should be separated as soon as possible and deep-frozen. Care is needed when the words 'deep-frozen' are mentioned outside the laboratory since this may result not infrequently in being presented with lysed whole blood owing to a misunderstanding. Combinations of antibiotics from patients on multiple therapy may mutually inactivate during transport. This is particularly true of carbenicillin and gentamicin where the high concentrations of the former can completely inactivate aminoglycoside activity during a day's transport at room temperature.

## Preparation of Agar Plates

The choice of an agar will have a direct bearing on the final result. Factors to consider are the physical qualities of the agar (strength when set, resistance to splitting if wells are to be cut), its pH and the quality of the resulting zone edges. Obviously it should support the growth of the indicator organism and not contain any inhibitors for the drug to be measured. The pH of the agar should ideally be the same as that of the standards and samples, or as near to it as possible. When assaying an antibiotic the activity of which is highly dependent upon pH, unpredictable effects can arise if the pH of the agar is different to that of the standards and samples.

To have an adequate number of replicates in an assay a dish of a minimum size is necessary. The effect of having the same number of replicates in a single dish or a number of smaller dishes is not the same because of inter-plate variation in the second case. Square assay dishes of 25 cm size are convenient for most assays as this area of agar can usually easily accommodate 30 doses (a dose being the application of a standard or sample to the plate) and is thus sufficient for 5 standards and 5 samples in triplicate. Disposable plastic dishes can be used (Ronsted, 1972) but they are expensive and distort if the agar is too hot when poured. They are useful if infected serum samples are anticipated. Stainless steel framed dishes with a sealed-in glass bottom may be re-used indefinitely with proper care. After an assay the agar is loosened and tipped out. The glass is ten scrubbed with a nail brush and hypochlorite scouring powder to remove any deposits of protein, and the dish thoroughly rinsed and dried. It can then be sterilized by u.v. irradiation in a safety cabinet. Autoclaving is unnecessary and may damage the dish.

Any variations in thickness will have a localized effect on zone size and thus whatever type of dish is used it is essential to have agar of absolutely uniform thickness which means levelling the bottom to a high degree of accuracy. This can be achieved by using an adjustable levelling device and a good quality spirit level, both of which must be in contact with the bottom of the dish itself. In the case of trays with metal edges extending under the bottom of the glass, to place the dish on a piece of level bench is not good practice, since the thickness of cement may vary from dish to dish, and from side to side of the same dish. Adjustable levelling devices can be bought or made. Alternatively 3 roundhead screws can be driven into a piece of otherwise unused bench and the heads accurately levelled.

## Indicator Organism

To maintain constancy of assay conditions it is essential to standardize the inoculum for each particular assay and to ensure that this inoculum is always used. When the organism can be maintained as a spore suspension this is relatively easy since suspensions will maintain their viable counts at 4°C over a period. Even then care should be taken to ensure constancy of agar temperature at spore addition because this may affect the final viable count. Occasionally spore suspensions become contaminated with vegetative organisms of the same or other strains. They can be removed by pasteurization, and the suspension is re-calibrated by viable counting.

Vegetative organisms can be maintained as a standard suspension in liquid nitrogen. Batches are prepared by growing in shaken liquid culture to the mid-logarithmic phase, adding glycerol to 10 per cent and freezing aliquots in individual ampoules. On thawing under standardized conditions each ampoule will give a consistent inoculum over a long period.

The indicator organism can either be incorporated into the agar before pouring the plate or can be flooded onto the agar surface after the agar has set and been well dried. Obviously the temperature of the agar is crucial when pouring incorporated plates containing a vegetative indicator organism. The use of a thermometer is advised if the occasional severe bacterial mortality is to be avoided. One advantage of an incorporated plate is that it can be surface sterilized by u.v. irradiation after dosing which tends to give clearer zone edges. Readable zones appear comparatively quicker with surface inoculated plates and are preferable for rapid assays.

## Application of Standards and Samples (doses) to the Plate

Doses may be applied by one of the following methods: Cut wells, dried paper discs, wet paper discs, cups and 'fish-spines'. The use of cups is now largely of historical interest. These cups and electricians porcelain 'fish-spines' both require careful cleaning before re-use. Wells and paper discs may be of any convenient diameter and, in the case of discs, the thickness can be varied. Wells are cut usually 'one at a time' in clinical laboratories by means of a sharp cork borer with a clean cutting edge. A machine to cut up to 64 wells simultaneously is available. A variety of methods are employed to lift out the agar discs. We find a fine pointed scalpel to pick out each disc as convenient as any. Whichever is employed it is important not to damage the edge of the well, and not to lift the main sheet of agar. This latter mishap can cause seepage of the well contents. Although this event can be prevented by sealing each well with a little molten agar, it is normally sufficiently uncommon not to warrant such a time-consuming process. The occurrence of seepage will be exposed by the failure of replicate wells to agree or by an unround zone. The wells should be 'filled' using a Pasteur pipette. By placing the tip of the pipette in the bottom of the well, gently squeezing the teat, and raising the pipette in one movement the well can be filled so that the liquid is level with the top of the agar with the meniscus flattened. If the well is filled drop-wise it is difficult to completely fill it, i.e., 4 drops may not fill it but the fifth drop may make it flow over the edge. An advantage in filling the wells to the brim instead of putting in a fixed volume is that should there be a slight variation in the thickness of the agar the amount of filling compensates to some extent for the irregularities. Provided the thickness of agar is constant either method is satisfactory at the level of accuracy required. For more precise assays fixed volume dosing and perfectly uniform agar are essential.

Paper discs (Whatman A.A.) are suitable for antibiotics that diffuse easily through the agar. The technique with these is to use fine pointed forceps to hold the disc whilst it is immersed in the antibiotic solution; several discs from the same sample may be dipped in turn and then laid on a sheet of dry filter paper for 10–20 s to allow excess fluid to drain. The forceps are then wiped and the discs may be applied directly to the surface of the agar with gentle pressure. The loading of discs in this manner is quick and gives results that are easily reproduced. There does not appear to be an advantage gained by measuring specific volumes onto them.

The choice of size of well or paper disc depends largely on the sensitivity required of the assay. The larger the physical diameter, the larger will be the zone for any particular concentration. When setting-up an assay, factors which may be varied by the choice of dosing method are sensitivity and differentiation in terms of zone size between adjacent standard concentrations. This latter factor is extremely important since one of the weakest points in plate assays is the actual measuring of zones. If adjacent standard concentrations have only a very small difference in zone size between them there is difficulty in discriminating between concentrations at that level. Assays should be arranged so that at what are clinically critical concentrations there is a relatively large change in zone size for a proportional change in concentration.

When dosing a plate, a template is essential to ensure the doses are put in their correct position. They should be spread sufficiently so that the zones of inhibition do not run into each other and so that they are not too near the edges of the plate where the thickness of agar

| Inadmissible Replication 30 Hole Plate | | | | | Random Code 30 Hole Plate | | | | |
|------|------|------|------|------|------|------|------|------|------|
| 20 | 10 | 5 | 2·5 | 1·25 | 1·25 | S2 | 2·5 | 10 | 5 |
| S1 | S2 | S3 | S4 | S5 | S1 | 20 | S3 | S4 | S5 |
| 20 | 10 | 5 | 2·5 | 1·25 | S3 | 5 | 1·25 | S2 | 10 |
| S1 | S2 | S3 | S4 | S5 | S4 | S1 | S5 | 20 | 2·5 |
| 20 | 10 | 5 | 25 | 1·25 | 2·5 | 10 | 5 | S4 | 1·25 |
| S1 | S2 | S3 | S4 | S5 | S2 | 20 | S1 | S5 | S3 |

**Fig. 2.7.** *Patterns of dosing for a 30-dose microbiological assay plate. Note that with inadmissible replication all the replicates of a standard (1·25–20) or a sample (S1-5) appear in the same position relative to the rest of the plate.*

may vary and the 'edge effect' can take place. This latter phenomenon is the increased zone size found at the edges of a plate compared to the same concentration away from the edge.

Consideration has to be given to the positions of the replicates of doses on the plates. These should be to a pre-determined 'random' pattern to avoid non-random influences affecting all the replicates of a dose in the same manner. For example, it would be unsatisfactory to have all the standards on the edge of a pattern and even worse to have them all on the same edge. Any slight unevenness in thickness of agar could then affect all the replicates similarly and a serious error could pass unnoticed. An example of inadmissible replication and a suitable random pattern are given in Fig. 2.7 for a 5 standard 15 unknown assay using 3 replicates. Randomization also helps to prevent bias occurring during reading when there is a subconscious urge to make replicates agree as nearly as possible. In our own laboratory the distribution of doses is given on a card used for dosing a plate, and reading and re-cording of zones is done sequentially before de-randomization. Several random patterns are available and are used in rotation in an effort to prevent staff from memorizing the patterns.

High precision assays often use patterns of dosing based on Latin square designs. In a true Latin square each dose (sample or standard) appears once in each row and each column. A typical design is shown in

Fig. 2.8. Because of the nature of the design each dose is replicated by as many times as the sum of the numbers of samples and standards. The sum of the zone size of each row and each column should be the same thus forming the basis for a convenient statistical analysis

```
3  4  8  1  7  6  5  2
2  7  3  8  5  1  4  6
8  1  5  2  4  3  6  7
4  2  1  6  3  5  7  8
7  8  2  5  6  4  3  1
6  5  7  3  8  2  1  4
1  3  6  4  2  7  8  5
5  6  4  7  1  8  2  3
```

**Fig. 2.8.** *A Latin square pattern of dosing for a microbiological assay plate. Each sample or standard appears in each column or row once, and the number of replicates equals the total number of standards and samples.*

and giving an indication of local deviations. Unfortunately this degree of replication is too great for clinical purposes and either a random pattern proved by controlled usage can be employed, as above, or some sort of quasi-Latin square design where each dose does not appear in all rows and columns (see Kavanagh, 1963).

## Pre-diffusion

The use of pre-diffusion depends on the antibiotic to be assayed and the indicator organisms used, and its value is found by experiment with any one particular assay. By pre-diffusing at room temperature or at 4°C for 2–3 h the zones sizes can be increased and, for example, we find this to be essential when trying to measure low concentration of carbenicillin with *Ps. aeruginosa* as the indicator organism.

## Incubation

The exact temperature of incubation is not critical and for most assays, except those employing *Micrococcus luteus*, 37°C is suitable, but what is most important is that the whole plate reaches the incubation temperature evenly. A large stack of assay plates should not be put into the incubator at one time since this will lead to uneven heating. It is better to put them in singly and then if necessary stack them when they are warm. The time of incubation will vary. Using a *Klebsielle* spp with surface inoculation it is possible to read good zones after $3\frac{1}{2}$–4 h and *B. subtilis*, staphylococci and *E. coli* can be read after 4–6 h, although it is generally easier to read the zones after overnight incubation. Using a good assay system there is no statistically significant difference between results from gentamicin assays read after 4 h and from overnight incubation.

## Measuring of Zones

Zone measuring is one of the weakest points of plate assays as performed in many routine laboratories. A ruler held against an assay plate can at the best measure to 0·5 mm. Using a pair of calipers measured on a ruler is not a lot better especially as the calipers may move before reading on ruler, introducing a further error. Vernier calipers are

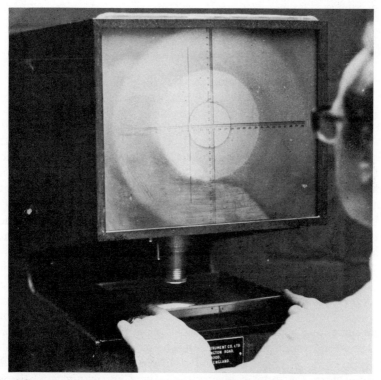

**Fig. 2.9.** *Optical manual zone reader in use. A good ergonomic position is maintained by using this instrument, thus minimising fatigue. The smaller circular image is the well cut in the agar and the larger is the zone itself.*

better as they do give a direct reading to nearest 0·1 mm. A backing sheet that has either a series of standard concentric circles or graduated diverging lines does speed up the reading of the zones, but these methods are not very accurate, as there are the inevitable problems of interpolation. The methods so far mentioned are all slow, ergonomically unsatisfactory and can lead to eye strain without the aid of magnification. There is also the added possibility of inaccurate observations because of either parallax or refraction errors when reading through the agar and plate.

For laboratories regularly performing antibiotic assays a magnifying manual zone reader is essential (Fig. 2.9). A bench-type reader which gives an illuminated image of the zones at about ×7 magnification makes it easy to read zones to the nearest millimetre with final images of 80–250 mm. Automated zone readers are available but are too expensive to be justified for use on the number of assays done in most clinical laboratories.

## Determination of Unknown Concentrations

After de-randomizing and matching the replicate zones, replicates are inspected for serious discrepancies and then meaned. Plotting a standard curve from which the unknown concentrations are read off, and calculation without graphic representation are the two principal methods by which unknown concentrations can be determined. The choice of a graphic plot will depend on which manipulation gives the nearest to straight line. For most assays a plot of log concentration against linear zone size will suffice. Occasionally a plot of log concentration against square of zone size will give the nearest fit a straight line. Whatever method is used a straight line fit cannot be assumed over a range of concentrations of four 2-fold dilutions. Points are joined with a Flexicurve or, less satisfactorily, by straight lines between points. Points which are patently discrepant are not included.

It is appropriate to digress a little at this point to consider how biological and technical variations affect the formation of the standard curve. An observed mean zone size for any particular concentration may only approximate to the theoretical true mean zone size for that concentration, the zone size for which will vary according to a normal distribution over a range (Fig. 2.10). The more replicates there are the more likely are the observed and true mean zones sizes to coincide. Furthermore, the more concentrations there are within a particular range the more observations will go to make up the standard curve as a whole. Clearly, using only two standard concentrations no informa-

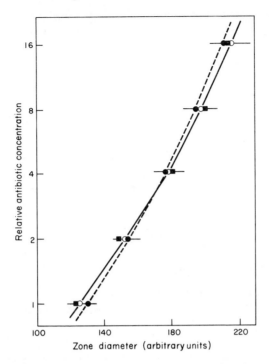

**Fig. 2.10.** *Hypothetical standard curves for microbiological assays. The theoretical correct mean zone size* (O) *for each standard concentration is given. The solid line joining these points is thus the theoretically correct standard curve. The hypothetical standard deviation for each mean is given:* -O-. *Because of random variations within this deviation the actual points which go to make up the standard curve are given:* ■. *Rarely these random variations may unfortunately compound to give a completely false standard curve:* -●--●-. *This figure illustrates the importance of having enough replicates for each point since the larger the numbers the smaller the chance of an abberant observed mean.*

tion is available at all about the linearity of a standard curve or the lack of it. With only 3 concentrations over the same range a significant error in one of the observed mean zone sizes will markedly influence the shape of the whole curve (Fig. 2.11). The same error occurring as one of 5 standard observations will have less influence on the overall shape of the standard curve, and it may be possible to exclude it altogether as an outlier.

Having plotted the observed mean zone sizes it will be necessary to draw a line of best fit and this can be a highly subjective process if the points do not fall on a smooth line. Removing this subjectivity is one of

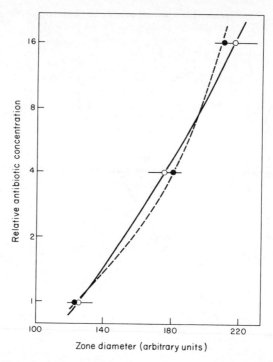

**Fig. 2.11.** *Hypothetical standard curve over same range as in Fig. 2.10 but with only three standards. The random deviations from the theretical means are similar to those in Fig. 2.10, but the smaller number of points has reduced the chances of the standard curve being as close to the theoretical one as in Fig. 2.10.*

the virtues of calculating a line of best fit and the method advocated by Bennett et al. (1966) will give a series of predicted zones for five standard concentrations arranged in 2-fold dilutions using a poly-nomial regression formula. A further extension to the method given by the authors is the calculation of the unknown concentrations from their observed mean zone sizes by the use of further formulae.

Although the mathematics are fundamentally simple, the manipula-tions themselves are time-consuming and access to a programmable calculator is essential if many unknowns are to be solved. A more sophisticated programmable calculator or a computer can be pro-grammed with random patterns of replicates in advance. Zones can be fed into the machine in the sequence as read and the whole operation of de-randomization, matching and meaning of replicates, and cal-culation of line of best fit and unknown concentration can be done as a single process. In our own department this procedure is done by a

Hewlett-Packard 9830A calculator. A typical print out is shown in Fig. 2.11. As a further safeguard both curves can be drawn on an X–Y plotter for visual confirmation of adequacy of fit.

## Control of Accuracy

*Record keeping and consistency of methodology.* The zone sizes and their means should be recorded in a standard format for every assay done. When assay methods are kept as technically consistent as possible any variation of zone size for a particular concentration of antibiotic will be an alert for a possible mishap. If assay conditions are not kept consistent then day to day variations will be large and serious error may go unnoticed.

*Internal controls.* Every assay plate should contain at least one internal control of known concentration which should be treated as an unknown sample. It is an advantage to have a variety of known concentrations labelled with known but random numbers, thus effectively preventing the assayist from being aware of the actual concentration before reference to a code. In the case of stable antibiotics this presents no problem since an internal control can be prepared broken into aliquots for storage and a fresh one used each day or week. For unstable antibiotics it may entail the preparation of a separate internal control from standard material on each occasion of assay.

The accumulated results for the internal standard will, when analysed, give some indication of the reproducibility of the assay method. To do this the percentage deviation of each observed value from the true (made-up) value can be calculated from

$$\frac{\text{observed} - \text{true}}{\text{true}} \times 100 \quad \text{per cent.}$$

The standard deviation of these observations will be an estimate of the reproducibility of the method, while the mean of the values might reveal any consistent under- or over-estimation (see Appendix 1 for method of calculation of 95 per cent confidence limits). It is important to remember that the accuracy as found by this method is as expressed over a number of assays performed on different occasions. There is no simple method of estimating the accuracy of any single assay plate.

Another form of control practised by some workers is to assay always a new batch of standards against the batch which is becoming outdated or is running out. If not exactly coincident, the two standard curves should at least be parallel. Finally, the validity of dilutions of samples

5 July 1975
Assay of Gentamicin                                    Plate No 17
Random Code 1.    30 Hole Plate

Zone sizes, as read from plate

| | | | | |
|---|---|---|---|---|
| 110 | 125 | 122 | 144 | 134 |
| 122 | 153 | 130 | 129 | 136 |
| 132 | 132 | 109 | 120 | 144 |
| 132 | 122 | 133 | 153 | 123 |
| 133 | 143 | 134 | 121 | 112 |
| 133 | 155 | 121 | 123 | 133 |

| | | Zone Sizes | | | Mean | mg/l |
|---|---|---|---|---|---|---|
| Std | 20·00 | 153 | 153 | 155 | 153·67 | 20·16 |
| | 10·00 | 144 | 144 | 143 | 143·67 | 9·87 |
| | 5·00 | 134 | 132 | 134 | 133·33 | 4·96 |
| | 2·50 | 122 | 123 | 123 | 122·67 | 2·55 |
| | 1·25 | 110 | 109 | 112 | 110·33 | 1·24 |
| Test | 1 | 122 | 122 | 121 | 121·67 | 2·40 |
| | 2 | 125 | 120 | 121 | 122·00 | 2·45 |
| | 3 | 130 | 132 | 133 | 131·67 | 4·46 |
| | 4 | 129 | 132 | 133 | 131·33 | 4·37 |
| | 5 | 136 | 133 | 133 | 134·00 | 5·18 |

Predicted Zones for Standards

| | | | | |
|---|---|---|---|---|
| 153·56 | 143·85 | 133·44 | 122·32 | 110·49 |

**Fig. 2.12.** *Typical print-out from Hewlett-Packard 9830A as programmed in our department to the mathematical method of Bennett et al., (1966). The zones as read are given first to enable any back-checking to be made easily. They are then presented derandomized and meaned. The predicted standard zones are calculated, and the results for the unknowns are calculated from this data and presented in the right-hand column. The mean zone sizes for the standards are resubstituted into the predicted curve and the results given above the values for the unknowns. This gives a quick check on the degree of deviation between the observed and predicted curves.*

should be checked where practical by ensuring that when corrected for dilution the same value is obtained as with an undiluted sample.

*External quality control.* While internal controls can ensure the reproducibility of a method it is essential to ensure the validity of results as compared with other estimating laboratories. If a laboratory's working standards and internal controls were prepared making the same incorrect assumption as to the potency of the initial standard material, the assay might appear to be correct while seriously under- or over-estimating the potency of samples. Such an error would be readily detected by estimating samples of known potency from an outside source circulated for this purpose. Experience of national surveys of quality control of gentamicin assays (Reeves & Bywater, 1975) have revealed that laboratories do have errors of this type. The more frequent error is, however, poor reproducibility, so much so that many laboratories would be unable to detect quite large shifts in the potency of standard material.

*Expression of results and S.I. units.* Système Internationale d'Unités (S.I. units) are now being used extensively in clinical chemistry. Where the molecular weight of the substance measured is accurately known it is recommended that results be expressed in molecular concentrations. In the case of antibiotics the molecular weight of what is being measured is not always known precisely. For example, gentamicin consists of at least 3 isomers which are present in slightly varying proportions. Furthermore, when pharmaceutical substances are being estimated which are normally administered by weight it makes sense to express their concentrations in mass concentration. The standard unit of volume in S.I. units is the litre and thus antibiotics should be expressed as mass per litre. This usually means mg/l at serum concentrations, and is no effort to convert from the more usual µg/l as it is the same figure numerically. The use of the litre is also an advantage when assays are done for pharmacological purposes since this unit of volume is the norm in pharmacokinetic investigations. All antimicrobial concentrations in this review are expressed in mg/l.

## Recommended Plate Assay Methods

Table 2.1 (pp. 54–55) gives the various conditions used in our laboratory for assaying the commonly encountered antimicrobial drugs. While the authors have found them all to be capable of producing results of adequate accuracy for clinical purposes, no claim is made that they are unique in this respect. The essential thing about any method is that it suits the purpose for which it is used, and that its accuracy is shown to be sufficient by suitable controls.

## Non-Microbiological Methods

Virtually all the chemical or physical methods available for measuring molecular substances have been applied to antimicrobial drugs. Unfortunately few of them are suitable for serum samples because of lack of sensitivity or because of susceptibility to interference the many constituents found in body fluids. For clinical purposes they also

**Table 2.1.** Laboratory conditions for the clinical assay of antimicrobial.

| Antimicrobial drug | Assay medium | Indicator organism | Inoculation of organism | | Dosing Wells (mm diam) | Discs (mm diam) | Range of Standards mg/l |
| --- | --- | --- | --- | --- | --- | --- | --- |
| | | | Incorporated | Surface | | | |
| Ampicillin and Amoxycillin | Penassay seed agar (Difco) | Bacillus subtilis (ATCC.6633) | + | – | – | 6 | 0·625–10 |
| Carbenicillin | Penassay seed agar (Difco) | Ps. aeruginosa (NCTC.10490) | – | + | 9 | – | 2–32 |
| Cephaloridine | Penassay seed agar (Difco) | Bacillus subtilis (ATCC.6633) | + | – | – | 6 | 0·5–8 |
| Clindamycin | Antibiotic medium No. 5 (Difco) | Staph. aureus (NCTC.6571) | + | – | 9 | – | 1·25–20 |
| Cloxacillin and Flucloxacillin | Penassay seed agar (Difco) | Staph. aureus (NCTC.6571) | + | – | 9 | – | 1·25–20 |
| Fusidic acid | see footnote* | Corynebacterium xerosis. (NCTC.9755) | + | – | 9 | – | 0·125–2 |
| 5-Fluorocytosine | Yeast morphology agar (pH 7·0) | Candida albicans | + | – | 9 | – | 1·25–20 |

| Antibiotic | Medium | Test organism | | | | | | |
|---|---|---|---|---|---|---|---|---|
| Gentamicin (clinical sample) | Diagnostic sensitivity test agar (Oxoid) | *Klebsiella* spp† ('Southmead strain') | − | + | − | 9 | − | 1·25–20 |
| Gentamicin (pharmacological sample) | Diagnostic sensitivity test agar (Oxoid) | *Bacillus pumilis* (NCIB.8982) | − | + | − | 9 | − | 0·625–10 |
| Kanamycin | Diagnostic sensitivity test agar (Oxoid) | *Klebsiella* spp† ('Southmead strain') | − | + | − | 9 | − | 1·25–20 |
| Lincomycin | Antibiotic medium no. 5 (Difco) | *Staph. aureus* (NCTC.6571) | + | − | + | 9 | − | 1·25–20 |
| Penicillin | Penassay seed agar (Difco) | *Bacillus subtilis* (ATCC.6633) | + | − | + | − | 6 | 0·625–10 |
| Streptomycin | Diagnostic sensitivity test agar (Oxoid) | *Klebsiella* spp† ('Southmead strain') | − | + | − | 9 | − | 2·0–32 |
| Tetracycline | Penassay seed agar (Difco) | *Staph. aureus* (NCTC.6571) | + | + | − | 7 | − | 0·625–10 |
| Tobramycin | Diagnostic sensitivity test agar (Oxoid) | *Klebsiella* spp† ('Southmead strain') | − | + | − | 9 | − | 1·25–20 |
| Trimethoprim | Diagnostic sensitivity test agar (Oxoid) 10 mg/l PABA | *E. coli* (NCTC.10418) | − | + | − | 9 | − | 0·625–10 |

\* The antibiotic assay of Fucidin in serum and other tissues. Leo Laboratories Ltd., Hayes, 1970.
† NCTC 10896 may be preferable because resistant to more antibiotics.

have the disadvantage that microbiologically inactive metabolites or breakdown products which may be measured are not necessarily relevant to therapy. Some methods also need expensive materials or equipment not often available in microbiological laboratories. For these reasons non-microbiological assays have not received much use in routine clinical laboratories to the present time. A number of methods, however, are regularly used and it is likely that others will be increasingly so as they are developed and the sophistication of micro-biology progresses.

Non-microbiological assay of sulphonamides by the Bratton and Marsahall (1939) procedure (see Varley, 1962) has been the standard method since shortly after their introduction. A colour reaction de-pendent on the same free amino group (N4) as microbiological activity ensures reasonably relevant specificity, although some metabolites, the proportion which vary between different sulphonamides, are

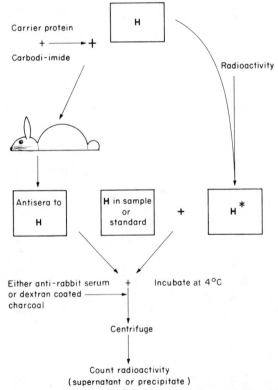

**Fig. 2.13.** *Schematic diagram to illustrate the principle of radioimmuno assay* H = *hapten (i.e. drug molecule).*

microbiologically inactive yet have a free N4 amino group. All metabolites may be assayed by prior hydrolysis with strong acid, while treatment with glucuronidase will allow the estimation of N4—glucuronides. Trimethoprim may be as conveniently assayed by a spectrofluorimetric method (Schwartz et al., 1969) as by microbiological assay, again with relevant specificity. A colorimetric assay for chloramphenicol is available (Levine & Fischbach, 1951; Grove & Randall, 1955) but a tedious extraction procedure is necessary if inactive metabolites are not to be measured.

Polarography is a classical tool of the physical chemist and is suitable for the estimation of certain antimicrobial agents. Working with protein-free ultra-filtrates of serum samples Benner (1971) used polarography to estimate penicillin and cephalosporins in the clinical or high-clinical range of concentrations. The result was available in two hours. Metronidazole a drug which may well receive wider use in the future for treating bacterial infection, can also be assayed by polarography Kane (1961). Some inactive metabolites in urine samples may be assayed (Ralph et al., 1974).

Spectrofluorimetry can offer sufficient sensitivity for the estimation of tetracyclines. A method for serum samples (Murthy & Goswami, 1973) based on the conversion of tetracycline to anhydrotetracycline (Hayes & DuBuy, 1964) gave a good recovery of drug.

Newer methods of non-microbiological assay have often involved the use of radio isotopes. Because not all microbiologists are familiar with isotopic technique two methods will be described in some detail and their characteristics discussed. It is not by chance that they are mainly directed at achieving a rapid and accurate assay for aminoglycosides (*see* Rapid Assays).

## Radio-immuno Assay (RIA)

Radio-immuno assay has been widely exploited for a number of years in the estimation of endogenous body components, for example, hormones, and a number of drugs. Its use for assaying antimicrobial drugs is a relatively late development. The principle of the assay (Fig. 2.11) is the measurement of the displacement by drug in the sample of radioactive drug added to the system which prevents it from binding with a specific antibody. After reaction of the antibody with the sample and radioactive drug, separation of the antibody-bound from the unbound drug is necessary. Drug/antibody complexes can be precipitated by the addition of a second antibody from a third species (for example, goat anti-rabbit globulin to precipitate rabbit anti-drug antibody).

The floccules are removed by centrifugation. Precipitation of the bound drug can also be achieved by physical agents such as organic solvents or inorganic sulphates. Alternatively, unbound drug may be selectively absorbed on to activated charcoal and removed by centrifugation. Either the radioactivity in the deposit or that remaining in the supernatant can be counted. A standard curve showing displacement of bound radioactivity by non-radioactive drug can be prepared from a set of standards and the unknown concentration determined (Fig. 2.14). An arithmetic plot of the curve is sigmoid but a degree of straightening can be achieved by various mathematical manipulations, for example a log–logit plot.

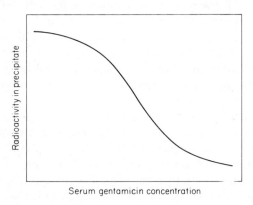

**Fig. 2.14.** *Typical, hypothetical standard curve for a radioimmuno assay of gentamicin. Increasing concentration of unknown drug displaces the radioactive drug from combination with the antibody and, hence, the precipitate.*

The type of counting equipment needed depends entirely on the label employed. At present aminoglycosides are not commercially available as labelled drugs. Labelling with $^3$H can be carried out by reaction with tritium oxide* but the resulting product contains only a small proportion of unchanged aminoglycoside. For some assays a degree of purification would be necessary. More promising is the use as an antigen of aminoglycoside linked to $^{125}$I by albumen. Many clinical chemistry laboratories have facilities for counting $^{125}$I, but few of them and even fewer microbiology laboratories have the liquid scintillation counter needed for $^3$H or $^{14}$C.

At present, there is no readily available source of the aminoglycoside antibodies also necessary for the assay. This again will present a problem to the smaller laboratory who might not have animal house facilities or the experience to prepare the antigen.

In spite of these disadvantages interest in the radio-immuno assay of aminoglycoside antibiotics, particularly gentamicin, has been manifest as the method promises a rapid and accurate assay (Lewis et al., 1972; Mahon et al., 1973; Berk et al., 1974; Lewis et al., 1975). These antibiotics are particularly suitable since under normal conditions virtually no metabolism occurs to inactive products which may react in the assay and give false values. The assay is specific for individual aminoglycosides and even for the individual isomers of gentamicin. The recent report of the use of $^{125}I$ linked to gentamicin is a particularly promising development as facilities for counting $^{125}I$ are available in most district hospitals and the gentamicin molecule itself would not need labelling (Watson et al., 1976; Broughton & Strong, 1976). At present the cost of any assay employing radio-activity is high compared with microbiological assays. For a general review of radio-immunoassay technique readers are referred to Sonksen (1974).

## Immunofluorescence Polarization Assay

This technique is under active development and appears to offer several advantages over conventional radio-immunoassay methods. No radio activity is required and, as there is no second antibody necessary to effect a separation of the $((^{125}I)$-antigen)—antibody complex from the supernatant, the complexity and the time necessary for the assay are reduced. The principle has been described in some detail (Udenfriend, 1962), and recently it has been developed for gentamicin determinations (Watson, et al., in preparation).

If a beam of polarized light is directed onto a small molecule either with native fluorescence, or complexed with a fluorescent tag, the emitted light will not be in the same plane of polarization as the excitation light due to the rapid Brownian rotation of the molecule. However, if a large molecule, e.g. an antibody protein, is attached to the small fluorescent molecule the rotational Brownian movement will be sufficiently showed down for an appreciable proportion of the emitted fluorescence to be in the same place of polarization as the excitation wavelength. Thus when an (antibody—(antigen–fluorescent tag)) complex is formed an increase in the fluorescence occurs which can be measured with a fluorimeter equipped with polarization filters in both the excitation and emission light-paths. The practical details for this technique remain to be accurately defined, but the method involves complexing gentamicin with a fluorescin label and then performing a standard competition-binding assay with gentamicin antiserum.

## Enzyme Radio-Tagging

Enzymes obtained from certain strains of bacteria resistant to aminoglycoside antibiotics are capable of producing inactivation by facilitating the transfer of a radical to the drug molecule. These include acetylating, adenylylating and phosphorylating enzymes. These radicals can be added to an assay system in a radioactive form, and their transfer to the drug molecule will in turn render radioactive the drug to be assayed (Fig. 2.15). Following separation from unreacted radical, the now radioactive drug is estimated by counting. Separation is conveniently done by putting the reactants on ion-exchange paper (Whatman P-81) which binds the highly basic aminoglycoside but allows the unreacted radical to be washed away. After drying the paper can then be counted in a liquid scintillation system.

The practical assay methods initially published used adenylylating enzyme (Smith et al., 1972; ten Krooden & Darrell, 1974; Phillips et al., 1974) but its affinity for ATP is comparatively low and high substrate concentrations were required to obtain sufficient counts above background for good accuracy within reasonable counting times. More recently acetylating enzymes have been exploited in practical assay methods (Haas & Davies, 1973; Broughall & Reeves, 1975a;

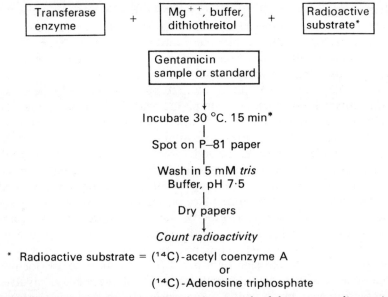

**Fig. 2.15.** *Schematic diagram to illustrate the principle of the enzyme radio-tagging assay of aminoglycoside antibiotics.*

Stevens et al., 1975). Because of higher affinity much higher counts can be obtained for equivalent substrate concentrations and these can be reduced to lower the unit cost of each reaction tube without impairing either speed or accuracy. The cost of reagents for a single gentamicin assay by one method is currently about 40p (Broughall & Reeves, 1975b) and of course, this reduces with the introduction of further tests. The time taken from receipt of sample to a result being obtained is 60 min.

The advantages of the radio-tagging method are its technical simplicity, the ease with which the reagents may be obtained and stored, together with their relatively low cost. The assay is specific to aminoglycoside antibiotics but they all mutually interfere in the assay (Fig. 2.16), except streptomycin. Mixtures of aminoglycosides antibiotics as therapy is fortunately a rare event. The main disadvantage of the assay is that the radicals used (acetyl or adenylyl) are labelled with β-emitters ($^{14}$C or $^3$H) and thus a liquid scintillation counter is required. Although in theory phosphorylating enzyme could be used for tagging with high-energy β-emitting phosphorus isotopes, they are particularly active and thus too dangerous to handle in the facilities offered by most pathology laboratories.

The characteristics of the various assays currently available for measuring gentamicin in serum are summarized in Table 2.2 (pp. 62–63).

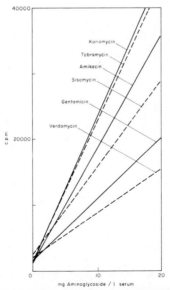

**Fig. 2.16.** *Typical standard curves for the acetyl transferase assay.*

**Table 2.2.** *Some practical methods of adequate sensitivity for assaying gentamicin in clinical samples.*

| Basic method | Variant | Receipt result time (h) | Consistently achieving ±25% accuracy | Specificity | Sample size (ml) | $^{14}$C or $^{3}$H genta. needed | Cost of material for single assay (pence) | Expensive, unusual equipment | Comments |
|---|---|---|---|---|---|---|---|---|---|
| Plate diffusion assay | standard[a] | 18 | Not easy but possible | Depends on indicator organism— lack always a hazard | 0·5 | No | <10 | None* | Require attention to detail to achieve good results. |
| | rapid[a] | 4-5 | | | | | | | |
| Vertical diffusion assay | —[b] | 18 | May be easier than plate assay. | Depends on indicator organism— lack always a hazard | 0·5 | No | < 5 | None | ? Deserving of greater attention. |
| Broth dilution assay | doubling | 18 | Impossible. | Depends on indicator organism— lack always a hazard | 0·5 | No | < 5 | None | Unacceptable method for gentamicin. |
| Urease assay | standard[c] | 1½ | Not easy but possible. | More specific than above methods | 1·0 | No | ? | Accurate pH meter. | Good results but tight internal control essential. |
| | semi-automated | ? | ? | | | | | Sampler ammonia electrode | Not yet fully developed. |

| Method | | Time (h) | Ease | Specificity | Sensitivity (µg/ml) | Anti-serum | No. | Equipment | Comments |
|---|---|---|---|---|---|---|---|---|---|
| Radio-immuno assay | ³H[d] | 2½ | Easy | Very high | <0.1 | Yes‡ | >50 | Liquid scintillation. Gamma counter. | ably not as convenient as acetyl transferase, unless very high specificity required for particular sample; ¹²⁵I-BSA looks promising. |
| | ¹²⁵I-BSA[l] | | | | | No | <50 | | No advantage over acetyl transferase. Most generally acceptable if access to counter. |
| Enzyme radio-tagging | adenylyl[f,g,h] transferase[l] | 2 | Less easy | Only other aminoglycosides interfere to varying degrees | <0.1 | No | 50–100 | Liquid scintillation counter | |
| | acetyl[i,j,k] transferase | 1 | Easy | | | | | | |
| Polarization immuno-assay | — | ½ ? | Easy | Very high | <0.1 | No‡ | ? <10 | Fluorimeter | Under development. May become method of choice, including automation. |

\* Zone reader considerable aid. as is access to computer for de-randomizing zones and calculating results.

† Can be shortened to 1 h.

‡ Requires anti-serum. not commercially available.

[a] Reeves (1972)
[b] Mitchison and Spicer (1949)
[c] Noone et al., (1971)
[d] Lewis et al., (1972)
[e] Berk et al. (1974)

[f] Smith et al., (1972)
[g] ten Krooden and Darrell (1974)
[h] Phillips et al., (1974)
[i] Haas & Davies (1973)
[j] Broughall & Reeves (1975a)

[k] Broughall & Reeves (1975b)
[l] Broughton & Strong (1976)
[m] Watson et al., (1976)
[n] Udenfriend (1962)
[o] Watson et al., (in preparation)

## Mixtures of Antimicrobial Agents

The problems presented to the clinical microbiology laboratory by mixtures of antimicrobial agents are the difficulties of measuring the concentration of one agent in the presence of others, often by relatively non-specific methods, and the unknown presence of interfering substances. This latter situation represents an obvious hazard to the accuracy of assays in samples from clinical sources.

*Methods available for assaying components of mixtures*
Specific assay
  (a) Chemical or physico-chemical method.
  (b) Radio-immuno assay.
  (c) Radio-tagging assay.
  (d) Immunofluorescence polarization assay.
Remove activity of unwanted component
  (a) Destruction.
  (b) Absorption.
  (c) Filtration.
  (d) Antagonism.
Separate components before assay
  (a) Electrophoresis.
  (b) Thin-layer chromatography.
  (c) Filtration.
Semi-specific microbiological assay
A wide variety of methods are available for assaying the components of mixtures, and are summarized above but some of them are not entirely suitable for use with serum samples. Some physical methods of separation often require de-proteinization before use. Because of the instability of many antibiotics strong acids cannot be used for this, and ultra-filtration or ultra-centrifugation is necessary to produce a protein-free sample. The method of assay must also have sufficient sensitivity to allow it to be used at the low concentrations often found in serum or other body fluids. Although a specific assay may seem ideal other methods are often more readily available or are just simpler.

### Specific Assays

Highly specific assays are not available for many antibiotics in common use but are still too numerous to detail here. Chemical or physico-chemical procedures are often needed and precautions to ensure that the method estimates only antimicrobially active drug in the context of its use are essential. Some of the methods, for example radio-immuno assay, are very specific, although some metabolites may be assayed (see p. 61). Others, for example the Bratton–Marshall procedure for estimating sulphonamide, will measure some metabolic products but not others. Microbiological assays can be rendered reasonably specific by using indicator organisms resistant to a number of antibiotics but sensitive to the one being assayed (see p. 70).

## Removal of Activity of Components before Assay

*Destruction of activity* by enzymatic means is widely used for penicillins and cephalosporins (Sabath et al., 1971; Stroy & Preston, 1971). Suitable broad-spectrum β-lactamase preparations are available commercially (e.g., Whatman β-lactamase broad-spectrum mixture; *see* Waterworth, 1973). As with all technical manoeuvres the enzymatic treatment should be closely controlled. In this instance the relevant β-lactam antibiotic in an appropriate concentration in serum should be treated simultaneously to the sample and assayed for loss of activity.

In theory other enzymes, usually from bacterial sources, such as gentamicin adenylylase, could be used for removing microbiological activity but unless commercial preparations are available it is unlikely that they would receive routine use. However they might be considered for a specific problem under carefully controlled conditions.

Less specific methods of destruction of activity can be tried. For example heating or changes in pH may inactivate less stable antiobiotics while stable drugs like aminoglycosides remain. Again a number of well designed controls would be essential. Ultraviolet light has been used to inactivate cephalothin (Sabath et al., 1968).

*Absorption of antibacterial drugs* onto activated charcoal or ion-exchange resins would be relatively non-specific. However, for a limited purpose such agents might be valuable. For example, the cation ex-

**Table 2.3.** *Absorption of antibiotics by Dowex 50-X8 Resin.*

| Completely absorbed | | Partially absorbed | | Not absorbed |
|---|---|---|---|---|
| Compound | max. conc. tested (mg/l) | Compound | max. conc. tested (mg/l) | compound |
| Kanamycin | 10 | Lincomycin | 20 | Benzyl penicillin |
| Gentamicin | 20 | Clindamycin | 20 | Ampicillin |
| Streptomycin | 50 | | | Cephaloridine |
| Tobramycin | 20 | | | Erythromycin |
| Neomycin | 100 | | | Doxycycline |

change resin (Dowex 50-X8) will remove basic antibiotics from serum (Reeves & Holt, 1976) (Table 2.3).

An interesting difference between the absorption of the aminoglycoside and the lincomycin is that the former seems to follow a first-order reaction while the latter reaches a limit of absorption (Fig. 2.17). The level at which this limit occurs depends on the proportion of resin to serum. The rate of absorption and, in the case of lincomycin, the final value is dependent on pH. The absorption is more efficient in buffer

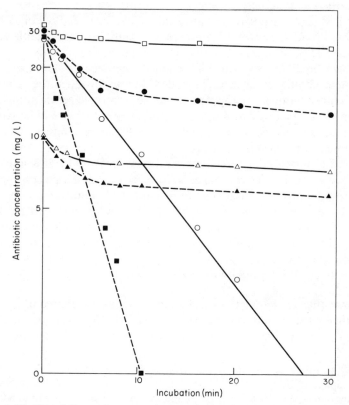

**Fig. 2.17.** *Use of the acidic ion-exchange resin Dowex 50-X8 to absorb two basic antibiotics, lincomycin and gentamicin. (Reeves & Holt, 1976).*

□——□  *lincomycin (starting conc. 30 mg/l). Resin 0·07 g/ml pH 8·4*
●——●  *lincomycin (starting conc. 30 mg/l). Resin 0·07 g/ml pH 6·7*
△——△  *lincomycin (starting conc. 10 mg/l). Resin 0·07 g/ml pH 7·6*
▲----▲  *lincomycin (starting conc. 10 mg/l). Resin 0·2 g/ml pH 7·6*
○——○  *gentamicin (starting conc. 30 mg/l). Resin 0·07 g/ml pH 8·4*
■----■  *gentamicin (starting conc. 30 mg/l). Resin 0·07 g/ml pH 6·7*

medium than in serum but aminoglycoside antibiotics can still be rapidly removed from the latter by Dowex 50-X8. As the resin is in large beads it settles out rapidly after agitation without centrifugation. The basic resin (Dowex 1X8-50) was not found to be successful in removing acidic antibiotic from serum although satisfactory absorption was achieved from buffer solutions.

*Differential molecular filtration* through collodion membranes has been described (Barza et al., 1973) but the degree of separation was not enough for clinical purposes. These authors and others (Wagman et al., 1975) discuss the problem of absorption of drug onto the filtration material.

*Antagonism of activity* without destruction of the drug molecule is well known for limited applications (Table 2.4). The activity of sulphon-amides may be inhibited by the addition of an excess of *para*-amino-benzoic acid to the growth medium, while trimethoprim is antagonized by thymidine. While again it must be emphasized that these manoeuvres be controlled to ensure the effectiveness of antagonism, we have found 50 mg/l of PABA and 10 mg/l of thymidine to be sufficient for assay of clinical samples. Semicarbazide has been used as an antagonist for streptomycin, and acid condition for erythromycin (Gove & Randall, 1955). Divalent cations have been used to inhibit aminoglycosides during the assay of clindamycin (Ervin & Bullock, 1974).

90cm

**Fig. 2.18.** *Equipment used for high-voltage electrophoresis of antibiotics. A water-cooled platten is housed in a plastic safety cabinet. Microswitches interrupt the current if the lid is raised. The electrodes of platinum wire on glass rods at each end are connected via the buffer boxes using lint wicks. (Stogate Instruments Ltd.).*

## Separation Before Assay

Separation of components of a mixture before assay has been widely practised for research and control applications in industry but little used in the clinical laboratory. *High-voltage electrophoresis* followed by bio-autography of antibiotics was originally described by Lightbown and de Rossi (1965) and a short report of its application to serum samples appeared in 1970 (Konno & Fujii). Following and modifying Lightbown and de Rossi's method we have been able to perform qualitative and quantitative resolutions of mixtures in serum samples from patients who have previously presented nearly insoluble problems (Reeves & Holt, 1975a). Electrophoresis is performed from sample wells in a non-nutrient buffered agarose layer of on a glass sheet resting on a water-cooled platten. Connections to boxes containing buffer are made by lint wicks and a voltage (up to 2000 V) is applied from a high capacity (up to 200 mA) power-pack (Fig. 2.18). The platten and boxes are contained in a safety cabinet whose lid activates microswitches to break the supply. Routinely, separation is conducted at 1200 V and 180 mA for 60 min. After separation the positions and concentration of the various antibiotics are detected by adding a seeded nutrient layer of agar in which zones appear after overnight incubation (Fig. 2.19). With suitable modifications the method has sensitivity sufficient for the concentration found in blood. Not all antibiotics have relative mobilities sufficiently different to enable a distinct separation of every possible combination to be made by the method as described, but various modifications (for example varying the pH of the running layer, the pre-treatment of samples with enzymes, and the use of antagonists in the seed layer) enable most problems to be overcome. With sufficient replication quantitation accurate enough for clinical purposes can be achieved.

**Table 2.4.** *Antagonism of antibacterial drugs.*

| Drug | Antagonist | Typical conc.* |
|------|------------|----------------|
| Sulphonamide | *para*-amino benzoic acid | 50 mg/l |
| Trimethoprim | thymidine | 10 mg/l |
| Streptomycin | semicarbazide | *see* Grove and Randall (1955) |
| Cycloserine | D-alanine | 20 mg/l |
| Aminoglycosides | lauryl sulphate or divalent cations | 20 g/l *see* Ervin and Bullock (1974) |

* For use in a suitable culture medium

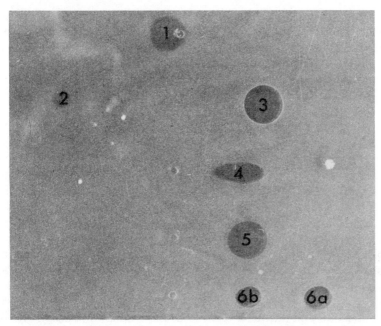

**Fig. 2.19.** *Demonstration separation of antibiotics by high-voltage electrophoresis and bio-autography. 1: cephaloridine 10 mg/l. 2: streptomycin 25 mg/l. 3: benzyl penicillin 5 mg/l. 4: cloxacillin 10 mg/l. 5: cephalothin 20 mg/l. 6(a): carbenicillin 20 mg/l. 6(b): benzyl penicillin as breakdown product of carbenicillin. The wells of origin are in vertical lines at the centre of the plate.*

Reports of the application of *thin-layer chromatography* followed by bio-autography to the separation of antibacterial drugs in clinical serum samples have been few. The problems are the influence of protein on the chromatography, and lack of sensitivity due to the relatively small volume of sample which can be applied. These strictures do not apply to investigations involving urine and there are numerous published examples of thin-layer chromatography for resolving mixtures of drugs or mixtures of parent drug and its metabolites. In spite of lack of sensitivity Konno and Fujii (1970) succeeded in using the method to separate and quantify penicillins from serum samples. More recently we have experimented with cellulose thin-layer chromatography as a means of separating some of the more commonly encountered antibiotics (Reeves & Holt, 1976). Appropriate technique has provided adequate sensitivity but separation has not always been achieved. For closely related antibiotics such as the aminoglycosides ion-exchange resins in thin-layers could be tried (Pauncz, 1972).

Thin-layer chromatography on Sephadex with bio-autography has been advocated as it involves the use of aqueous rather than organic solvent (Zuidweg et al., 1969). Experience with the method has not produced sufficient sensitivity to allow typical serum samples to be assayed (Reeves & Holt, 1976).

## Semi-specific Microbiological Assay

A semi-specific microbiological assay by *differential microbial resistance* can often provide a practical solution to the assay of one antibiotic in the presence of another. The real problem of this method is that to cover the possible contaminants a large number of indicator strains need to be kept. Each will have its own characteristics in assay and it is likely that the occasional assay with an unfamiliar organism would be inaccurate. Sometimes naturally resistant organisms are not available and the required resistance can be produced by training on antibiotic strains produced might well have undesirable characteristics such as slow growth in the first instance and spontaneous reversion to sensitivity in the second. Such strains therefore require careful maintenance and adequate control in use. Examples of assays by differentially resistant organisms are too numerous to be given here but they vary from the use of a multiply-resistant *Providentia stuartii* used to assay amikacin (Marengo et al., 1974) to the assay of gentamicin in the presence of lincomycin by using the natural resistance of a coliform as indicator organism. An anaerobic assay for clindamycin in presence of gentamicin using *Clostridium perfringens* (Sabath & Toftegaard, 1974), and an assay for gentamicin in presence of cephalothin, ampicillin, carbenicillin, clindamycin and chloramphenicol using *Staph. epidermidis* (Alcid & Seligman, 1973), have both been described recently. As an alternative to a resistant indicator organism, special conditions in the assay may render it partially specific. For example, using a medium at pH 6·0 will effectively reduce aminoglycoside activity. In using all the above types of assay it is important to check that not only is the indicator strain resistant to the unwanted drug(s) alone but that it remains fully resistant when the drug to be assayed is also present.

Other methods of differential agar diffusion assay rely on the indicator organism being substantially less sensitive to the interfering antibiotic than the one to be assayed. For example, penicillin at typical blood levels can be assayed using *Sarcina lutea* in the presence of streptomycin provided the concentration of the latter does not exceed 80 mg/l (Raahave, 1974). In other circumstances the indicator strain may be as sensitive to the interfering drug as to the assayed drug but the large ratio of assayed over interfering drug concentrations in typical body fluid samples permits an accurate assay. For example, in patients on typical doses of carbenicillin and gentamicin the concentrations of the former may exceed the latter a hundredfold. Assay using a sensitive *Pesudomonas aeruginosa* is possible because the necessary dilution of samples lowers the activity of gentamicin to a level where it no longer interferes in the assay.

Finally, when other methods have failed it is possible to assay one drug (A) in the presence of another (B) making an allowance for the activity of B. It is necessary to specifically assay B which is then added to the A standards in the same concentration as the sample to be assayed for A. This procedure involves a great deal of work and is lengthy, therefore not recommending itself for use in clinical laboratories. The complexity of such assays are illustrated by Lightbown (1970) and O'Callaghan et al., (1968).

## Undisclosed Contamination of Samples

The unknown presence of antibacterial agents other than that to be assayed may well cause an inaccurate result due to the relative non-specificity of many methods. In our experience this situation occurs with alarming frequency (Reeves & Holt, 1975). Screening by high-voltage electrophoresis and bio-autography we found nearly 20 per cent of 189 consecutive sera submitted for assay contained agents undisclosed with the request. These were confirmed by a restrospective examination of the patient's records which also revealed that this apparent lack of communication was due most often to a recent cessation of treatment with the contaminating drug. Measurable quantities of anti-bacterial drugs persist for some 24 h after the last dose, and for much longer when renal function is impaired. Other causes were oversight by the clinicians, failures of communication between clinicians, or a lack of appreciation of the non-specificity of the assay method. Uncharacteristic zones (e.g., a hazy edge) for a sample, or a pharmocologically unlikely or impossible result may alert the laboratory to the presence of undisclosed drugs. Whenever this happens it always makes one apprehensive of how many times it has occurred to a lesser extent and passed undetected.

## Rapid Assays

Although methods of optimizing the dosage of potentially toxic anti-

D

biotics such as aminoglycosides exist, unpredictable pharmacology or caution because of overdosing may result in less-than-ideal peak blood levels. Furthermore, all dosage calculations depend on estimates of renal function and as this changes rapidly in seriously ill patients, the lag between change in serum-creatinine or -urea after a change in renal function is important. Although changes in function may be reflected in the urine, collections of urine may be difficult. Failure to achieve sufficiently high serum levels is undesirable in the seriously ill patient and a rapid assay with the result available before the next dose is due would be advantageous. In practice the dose interval is not less than 8 h so an assay taking about four hours to produce a result is adequate, making allowances for taking the sample 1h after the dose and its transport. Even more rapid assays also have a convenience factor in that the assay may be requested at any time and the shorter the receipt–result interval, the more likely it is to be completed in the working day.

Conventional microbiological agar diffusion assays have incubation periods of 18 h or more. To qualify as a rapid assay for clinical use the incubation period would have to be shortened to 4–5 h. An assay for aminoglycosides using plates pre-seeded with *Bacillus subtilis* and stored at 4°C has been described by Sabath et al. (1971), the authors claiming a readable result in 2–4 h. Similar assays using *Staph. aureus* (Warren et al., 1972), *Staph. epidermidis* (Alcid & Seligman 1973) and *Bacillus subtilis* (Justesen, 1973) have also been reported. With the exception of the strain of *Staph. epidermidis* all the indicator organisms used were sensitive to clindamycin and lincomycin, one of which is often used in combination with an aminoglycoside. Furthermore, although interference by penicillins and cephalosporins can be overcome with β-lactamase, unless β-lactamase is used with every sample, the assay is susceptible to inaccuracy due to the unknown presence of these interfering drugs. For these reasons a coliform indicator organism is preferred by many laboratories when assaying aminoglycosides in samples from patients. A *Klebsiella* spp. can be surface inoculated (Reeves, 1972) or incorporated (Lund et al., 1973) and the assay read after 3–4 h incubation.

Rapid plate diffusion assays are as accurate as the corresponding conventional method (Warren et al., 1972; Lund et al., 1973). From our own, data a comparison of the results of 84 serum samples ranging in gentamicin concentrations from 0·4 mg/l to 15·4 mg/l assayed by the 18 h and 4 h methods gave a correlation coefficient of 0·995.

Turbidimetric assays can produce rapid results but, for reasons discussed previously, have not been developed for clinical use to any great extent. The urease method has also been mentioned already (see p. 36).

Methods utilizing the inhibition by aminoglycosides of the pH change produced by the bacterial metabolism of carbohydrate (Fain & Knight, 1968; de Louvois, 1974) have been reported. Although a rapid result is obtainable the authors do not give sufficient data on which a proper assessment of accuracy can be made. A rapid cephalosporin assay using the same principle has been described by Noone (1973).

When available for the antimicrobial agent under consideration, chemical or physico-chemical methods usually give rapid answers and are discussed elsewhere in this review. Unfortunately they often require reagents or equipment unavailable to the clinical microbiologist.

## Conclusions

With the increasing awareness of individual differences in the pharmacology of drugs between patients, the importance of measuring the blood levels of drugs in clinical practice has recently been emphasized. Thus the need for readily and accurately measuring antibiotic concentrations is real and nowhere is this more important than in the management of the patient seriously ill with bacterial infection. Much of the principle of and detail of these measurements has been set out in this chapter. If there is a message to be taken by the reader to the laboratory it is one of 'method for method's sake' not being important. What is essential is that the method chosen works well for the samples taken in their particular circumstances and in the hands of the laboratory worker concerned. Furthermore, it must be shown to work well by adequate internal and external controls. Choose wisely and work carefully.

## Appendix 1. Method of Calculation of 95 per cent Confidence Limits of Assay from Internal Controls

The accuracy of a method can be assessed by assaying one or more internal controls over a period of time. The deviation of the result of each assay of a control is expressed as a percentage difference, $x$

$$x = \frac{(\text{Observed value} - \text{true value})}{(\text{True value})} \times 100$$

The mean of the percentage differences ($\bar{x}$) is found by:

$$(\bar{x}) = \frac{\Sigma(\bar{x})}{n}$$

where $n$ is the number of assays performed.

The standard deviation, $s$ is found by:

$$(s) = \frac{(\Sigma x^2 - (\Sigma x)^2 n)}{(n - 1)}$$

The 95 per cent confidence limits (i.e., those limits for percentage error within which 19 out of every 20 assay results should fall) are calculated by:

$$\text{95 per cent limits} = \bar{x} + ts,$$

where $t$ is the value for $n$ found in tables of the $t$-distribution:

$$t = 2{\cdot}8 \text{ for } n = 5$$

$$t = 2{\cdot}3 \text{ for } n = 10$$

$$t = 2{\cdot}0 \text{ for } n = >20$$

## Appendix 2. Manufacturers of Equipment and Materials Cited

Autodata, *80, Walsworth Road, Hitchin, Herts. SG4 9SX. Telephone: Hitchin 3823.*

Difco Laboratories, *P.O. Box 14B, Central Avenue, West Molesey, Surrey, KT8 0SE, Telephone: 01-979 9951.*

Leebrook Instrument Co., Ltd., *Sarum Farm, Sarum Road, Winchester, Hants. Telephone: 0962–3823.*

Mast Laboratories Ltd., *38, Queensland Street, Liverpool, L7 3JG. Telephone: 051–709 9826–8.*

Nunc, U. K. Agent, *Jobling Lab. Division, Stone, Staffs, ST15 OBG. Telephone: 0785–83 2121.*

Oxoid Ltd., *Wade Road, Basingstoke, Hants RG24 0PW. Telephone 0256–61144.*

Sigma London Chemical Co. Ltd., *Norbiton Station Yard, Kingston-upon-Thames, Surrey, KT2 7BH.*

Stogate Technical Developments Ltd., *Industrial Estate, Daux Road, Billinghurst, Sussex, RH 14955.*

Whatman Biochemicals Ltd., *Springfield Mill, Maidstone, Kent, ME14 2LE.*

# Further Reading

## *Absorption, Distribution and Elimination of Antibiotics*

Goldstein, A., Aronow, L. & Kalman, S. M. (1974) Principles of Drug Action. New York: Wiley Interscience.

La Du, B. N., Mandel, H. G. & Way, E. L. (1971) *Fundamentals of Drug Metabolism and Drug Disposition.* Baltimore: Williams & Wilkins Co.

Smith, S. E. & Rawlins, M. D. (1973) *Variability in Human Drug Response.* London: Butterworths.

Cadwallader, D. E. (1973) *Biopharmaceutics and Drug Interactions,* Basle: Hoffman-La Roche.

## *Assay Methods*

Blair, J. E., Lennette, E. H. & Truant, J. P. (1974) *Manual of Clinical Microbiology.* 2nd ed. Baltimore: *Am. Soc. Microbiol.*

Garrod, L. P., Lambert, H. P. & O'Grady, F. (1973) *Antibiotic and Chemotherapy.* 4th ed. London: Churchill Livingstone.

Grove, D. C. & Randall, W. A. (1955) *Assay Methods of Antibiotics.* New York: Medical Encyclopedia.

Kavanagh, F. (1963) *Analytical Microbiology.* New York: Academic Press.

Kavanagh, F. (Ed.) (1972) *Analytical Microbiology.* Vol. II, New York: Academic Press.

Sonksen, P. H. (Ed.) (1974) Radioimmunoassay and Saturation analysis. *Br. med. Bull.,* **30,**

# References

Alcid, D. V. & Seligman, S. J. (1973) Simplified assay for gentamicin in the presence of other antibiotics. *Antimicrob. Ag. Chemother.,* **3,** 559.

Barza, M., Bergeron, M. G., Brusch, J. & Weinstein, L. (1973) Selective filtration of antibiotics through collodion membranes. *Antimicrob. Ag. Chemother.,* **4,** 337.

Benner, E. J. (1971) Two-hour assay for content of penicillins and cephalosporins in serum. *Antimicrob. Ag. Chemother.*—1970. 201.

Bennett, J. V., Brodie, J. L., Benner, E. J. & Kirby, W. M. M. (1966) Simplified, accurate method for antibiotic assay of clinical specimens. *Appl. Microbiol.*, **14**, 170.

Berk, L. S., Lewis, J. L. & Nelson, J. C. (1974) One-hour radioimmunoassay of serum drug concentrations as exemplified by digoxin and gentamicin. *Clin. Chem.* **20**, 1159.

Blair, J. E., Lennetti, E. H. & Truant, J. P. (1974) *Manual of Clinical Microbiology*, 2nd ed, Baltimore: *Am. Soc. Microbiol.*

Bratton, A. C. & Marshall, E. K. (1939) A new coupling method for sulfanilamide determination. *J. biol. Chem.* **128**, 537.

Broughall, J. M. & Reeves, D. S. (1975a) The acetyltransferase enzyme method for the assay of serum gentamicin concentrations and a comparison with other methods. *Journal clin. Path.*, **28**, 140.

Broughall, J. M. & Reeves, D. S. (1975b) Properties of the gentamicin acetyltransferase enzyme and application to the assay of aminoglycoside antibiotics. *Antimicrob. Ag. Chemother.*, **8**, 222.

Broughton, A. & Strong, J. E. (1976) Radioimmunoassay of iodinated gentamicin. *Clin. Chim. Acta*, **66**, 125.

de Louvois, J. (1974) A rapid method of assaying gentamicin and kanamycin concentrations in serum. *J. med. Microbiol.*, **7**, 11.

Elfving, J. & Pettay, O. (1974) Some sources of error in monitoring antibiotic treatment. Progress in Chemotherapy, Ed. Daikos, Hellenic Society for Chemotherapy. pp. 211–218,

Ervin, F. R. & Bullock, W. E. (1974) Simple assay for clindamycin in the presence of aminoglycosides. *Antimicrob. Ag. Chemother.* **6**, 831.

Fain, S. & Knight, D. C. (1968) Rapid microbiological assay of antibiotics in blood and other fluids. *Lancet*, **2**, 375.

Garrod, L. P., Lambert, H. P. & O'Grady, F. (1974) *Antibiotic and Chemotherapy*, 4th ed. London: Churchill Livingstone.

Grove, D. C. & Randall, W. A. (1955) Assay Methods of Antibiotics. *A Laboratory Manual*. New York: Medical Encyclopaedia.

Haas, M. J. & Davies, J. (1973) Enzymatic acetylation as a means of determining serum aminoglycoside concentrations. *Antimicrob. Ag. Chemother.*, **4**, 497.

Hayes, J. E. & Du Buy, H. G. (1964) A simple method for quantitative estimation of tetracycline antibiotics. *Anal. Biochem.*, **7**, 322.

Justesen, T. (1973) A rapid micro-assay of gentamicin, kanamycin, or streptomycin in serum containing penicillins or cephalosporins. *Acta Path. Microbiol. Scand. (Section B)*, **81**, suppl. 241. 111.

Kane, P. O. (1961). Polarographic methods for the determination of two antiprotozoal nitroimidazole derivatives in materials of biological and non-biological origins. *J. polarographic. Soc.*, **7**, 58.

Kavanagh, F. (1963) *Analytical Microbiology*. New York: Academic Press.

Kavanagh, F. (1972a) *Analytical Microbiology*, Vol. II, New York: Academic Press.

Kavanagh, F. (1972b) Photometric assaying. *In: Analytical Microbiology*, Vol. II, 44–121, New York: Academic Press.

Konno, M. & Fujii, R. (1970) Determination of separate antibiotic levels by a chromotographic assay method in the body fluid following medication of combination of antibiotics. *Progress in Antimicrobial and Anticancer Chemotherapy*, Vol. **1**, Baltimore University Park Press, 1027–1032.

La Du, B. N., Mandel, H. G. & Way, E. L. (eds) (1971) *Fundamentals of Drug Metabolism and Drug Disposition.* Baltimore: Williams & Wilkins, Co.

Lancet (1974) Serum-gentamicin, *Lancet,* **2,** 1185.

Levine, J. & Fischbach, H. (1951). The chemical determination of chloramphenicol in biological materials. *Antibiot. Chemother.* **1,** 59.

Lewis, J. E., Nelson, J. C. & Elder, H. A. (1972) Radioimmuno assay of an antibiotic: gentamicin. *Nature, New Biology,* **239,** 214.

Lewis, J. E., Nelson, J. C. & Elder, H. A. (1975) Amikacin: a rapid and sensitive radioimmunoassay. *Antimicrob. Ag. Chemother.* **7,** 42.

Lightbown, J. W. (1970) Assay of individual antibiotics in drug combinations. *In: The Control of Chemotherapy,* pp., 1–9, Watt, P. J., Ed. Edinburgh: Livingstone.

Lightbown, J. W. & de Rossi, P. (1965) The identification and assay of mixtures of antibiotics by electrophoresis in agar gel. *Analyst,* **90,** 89.

Lund, M., Blazevic, D. J. & Matsen, J. M. (1973) Rapid gentamicin bioassay using a multiple-antibiotic-resistant strain of *Klebsiella pneumoniae. Antimicrob. Ag. Chemother.* **4,** 569.

Mahon, W. A., Ezer, J. & Wilson, T. W. (1973) Radioimmunoassay for measurement of gentamicin in blood. *Antimicrob. Ag. Chemother.* **3,** 585.

Marengo, P. B., Wilkins, J. & Overturf, G. D. (1974) Rapid, specific microbiological assay for amikacin (BB-K8). *Antimicrob. Ag. Chemother.,* **6,** 498.

Mitchison, D. A., and Spicer, C. C. (1949) A method of estimating streptomycin in serum and other body fluids by diffusion through agar enclosed in glass tubes. *J. gen. Microbiol.* **3,** 184.

Murthy, V. V. & Goswami, S. L. (1973) A modified fluorimetric procedure for the rapid estimation of oxytetracycline in blood. *J. clin. Path.,* **26,** 548.

Noone, P. (1973) A rapid assay method for cephalosporins. *J. clin. Path.,* **26,** 506.

Noone, P., Pattison, J. R. & Slack, R. C. B. (1971) Rapid assay of gentamicin. *Lancet,* **2,** 1194.

Noone, P., Pattison, J. R. & Slack, R. C. B. (1975) A simple, rapid assay for the measurement of antibiotic concentrations in human serum. *In: Some Methods for Microbiological Assay.,* pp. 153–167. Board R. G. & Lovelock, D. W., Eds London: Academic Press.

O'Callaghan, C. H., Kirby, S. M. & Wishart, D. R. (1968) Microbiological assay of mixed cephalosporins. *Antimicrob. Ag. Chemother.,* 716–722.

Pauncz, J. K. (1972) Thin-layer chromatography of basic water soluble antibiotics on resin-coated chromoatoplates. *J. Antibiotics,* **25,** 677–678.

Phillips, I., Warren, C. & Smith, S. E. (1974) Serum gentamicin assay: a comparison and assessment of different methods. *J. clin. Path.,* **27,** 447.

Raahave, D. (1974) Paper disk-agar diffusion assay of penicillin in the presence of streptomycin. *Antimicrob. Ag. Chemother.,* **6,** 603.

Ralph, E. D., Clarke, J. T., Libke, R. D., Luthy, R. P., and Kirby W. M. M. (1974) Pharmacokinetics of metronidazole as determined by bioassay. *Antimicrob. Ag. Chemother.* **6,** 691.

Reeves, D. S. (1972) Assay of gentamicin. *Lancet,* **2,** 1369.

Reeves, D. S. (1974) Gentamicin therapy. *Br. J. hosp. Med.,* 837–850.

Reeves, D. S. (1975) Laboratory and clinical studies with sulfametopyrazine as treatment for bacteriuria in pregnancy. *J. antimcrob. Chemother.,* **1,** 171.

Reeves, D. S. & Bywater, M. J. (1975) Quality control of serum gentamicin assays—experience of national surveys. *J. antimicrob. Chemother.,* **1,** 103.

Reeves, D. S. & Holt, H. A. (1975) Resolution of antibiotic mixtures in serum samples by high-voltage electrophoresis. *J. clin. Path.*, **28**, 435.

Reeves, D. S. & Holt, H. A. (1975b) Laboratory problems from antibiotic combinations *Proc. IXth. Internat. Cong. Chemother.*, 1975.

Ronsted, P. (1972) Disposable plastic tray for large plate assays of antibiotics. *Antimicrob. Ag. Chemother.*, **2**, 49.

Sabath, L. D., Loder, P. B., Gerstein, D. A. & Finland, M. (1968) Measurement of three antibiotics (penicillin, cephalothin and chloramphenicol) when present together in mixtures. *Appl. Microbiol.*, **16**, 877.

Sabath, L. D., Casey, J. I., Ruch, P. A., Stumpf, L. L. & Finland, M (1971) Rapid microassay of gentamicin, kanamycin, neomycin, streptomycin and vancomycin in serum or plasma. *J. Lab. clin. Med.*, **78**, 457.

Sabath, L. D. & Toftegaard, I. (1974) Rapid microassays for clindamycin and gentamicin when present together and the effect of pH and each on the antibacterial activity of the other. *Antimicrob. Ag. Chemother.*, **6**, 54.

Schwartz, D. E., Kochelin, B. A. & Weinfeld, R. E. (1969). Spectrofluorometric method for the determination of trimethoprim in body fluids. *Chemother.*, **14**, 22.

Smith, D. H., van Otto, B. & Smith, A. L. (1972) A rapid chemical assay for gentamicin. *New Engl. J. Med.*, **286**, 583.

Sonkren, P. E., Ed. (1974) Radio-immunoarray and saturation analysis. *Br. Med. Bull.*, **30**.

Stevens, P., Young, L. S. & Hewitt, W. L. (1975) Improved acetylating radioenzymatic assay of amikacin, tobramycin, and sisomicin in serum. *Antimicrob. Ag. Chemother.***7**, 374.

Stroy, S. A. & Preston, D. A. (1971). Specific assay of aminoglycosidic—or polymyxin—type antibiotics present in human serum in combination with cephalosporins. *Appl. Microbiol.* **21**, 1002.

ten Krooden, E. & Darrell, J. H. (1974) Rapid gentamicin assay by enzymatic adenylylation. *J. Clin. Path.* **27**, 452.

Undenfriend, S. (1962) *Fluorescence Assay in Biology and Medicine.* pp. 197–8, New York: Academic Press.

Varley, H. (1962) *Practical Clinical Chemistry*, p 632, London: Heinemann.

Wagman, G. H., Bailey, J. V., Weinstein, M. J. (1975). Binding of aminoglycoside antibiotics to filtration materials. Antimicrobial Agents and Chemotherapy, **7**, 316.

Warren, E., Snyder, R. J. & Washington, J. A. (1972) Four hour microbiological assay of gentamicin in serum. *Antimicrob. Ag. Chemother.*, **1**, 46.

Waterworth, P. M. (1973). An enzyme preparation inactivating all penicillins and cephalosporins. *J. Clin. Path.*, **26**, 596.

Watson, R. A. A., Landon, J., Smith, D. S. & Shaw, E. J. (1976) Polarisation fluoroimmunoassay of gentamicin. (in preparation).

Watson, R. A. A., Shaw, E. J., Edwards, C. R. W. & Landon, J. (1976) An I[125] Based Radioimmunoassay for Serum Gentamicin. *Proc. IXth Internat. Cong. Chermother.* July, 1975.

Zuidweg, M. H. G., Ostendorf, J. G. & Bos, C. J. K. (1969) Thin layer chromatography on Sephadix for the identification of antibiotics. *J. Chromatography*, **42**, 552.

# 3 | Inhalation Therapy Equipment

JAY P. SANFORD and ALAN K. PIERCE

## Hazards of Pulmonary Infection Associated with Ventilators

Almost since the introduction of anesthesia, concern regarding the care and disinfection of anesthesia equipment has been expressed. Skinner is quoted as having written in 1873:

> If there be one evil more crying, more disgusting than another, in the practice of inducing anesthesia, it is the use of inhalers . . . There is not one inhaler, my own excepted, where every patient is not made to breathe through the same mouthpiece, tube, and chamber . . . Sweet seventeen is made to follow a bearded devotee to Bacchus, saturated with the smoke of cigars and the exhalations of cognac . . .

and

> . . . speak of refinement! We turn up our noses if we have not a clean table-napkin every day, if our knife, fork, spoon and plate be not cleaned or changed after every dish . . . but when we come to inhalation . . . after twenty-five years' experience . . . we remain the merest barbarians, everyone breathing after his neighbor . . . (Duncom, 1947).

Much later, concern for the prevention of cross infection from anesthesia machines began to appear in the literature (Livingston et al., 1941). From a series of experiments, Magath estimated that 'during an hour's anesthesia, an average of 120 bacteria find their way into the machine' (Lundy, 1943). He also commented that:

> It is by no means certain that they may contaminate the side of the machine through which inspiration occurs. But the possibility of even a few tubercle bacilli, diphtheria organisms, or virulent pneumococci being transmitted in such a manner to a patient is not to be tolerated.

For these reasons, Magath devised a watertrap for filtering out bacteria, noting, 'It is a well-known fact that bacteria in water will not be transmitted unless the water is sprayed into the air'. This observation becomes of major importance in the later discussion on inhalation therapy equipment. At approximately the same time, Adriani and Rovenstine wrote that, 'There are many reports of experiments demonstrating that absorbing units permit nebulized suspensions of bacteria to pass the soda. These results do not furnish evidence that cross infection of one patient to another occurs under clinical conditions' (Adriani & Rovenstine, 1941). These investigators then nebulized broth suspensions of cultures of viable *Escherichia coli* and *Mycobacterium tuberculosis* into canisters. While they had evidence that bacteria had passed into the canister, the plates exposed to air currents simulating inhalations and exhalations showed no growth, from which they concluded that 'the possibility of cross-infection, by using a canister for one patient and immediately for another, is remote'. These observations have been quoted almost in full since they both illustrate and define many of the current areas of problems and confusion 30 years later.

Following the Second World War, the principles of respiratory physiology were applied to development of ventilatory support procedures. Hoffman and Finberg probably were the first investigators to record an increased occurrence of infection due to *Pseudomonas aeruginosa* coincident with the use of aerosol equipment (Hoffman & Finberg 1955). They observed that following the introduction of nebulizers to create a high humidity environment for infants in incubators, there was an increased incidence of bacteremia, dermatitis, conjunctivitis and omphalitis due to pseudomonads. They were not able to isolate organisms from the reservoir jars of nebulizers or incubator walls (a finding which is difficult to reconcile in the light of current knowledge), and hence implicated the moist environment rather than nebulization per se as the basis of their observations. Macpherson (1958) and Sever (1959) next called attention to inhalation therapy as a potential vector of hospital-associated pneumonia when each reported that the water found in humidifying devices was regularly contaminated with large numbers of a variety of bacteria. Bishop and associates and Phillips and Spencer suggested the potential danger of mechanical ventilators (Bishop et al., 1962, 1963, 1966; Phillips & Spencer, 1965). At approximately the same time, Sykes and our group reported on means of ventilatory sterilization and disinfection (Sykes, 1964; Reinarz et al., 1965).

There now exists a general awareness of the potential role of inhalation therapy equipment and practices as vectors in the causation of hospital-associated pneumonia. However, one of the problems which remains to be better defined is the magnitude of the risk of hospital-associated pneumonia which can be linked to the use of respiratory support equipment. While pseudomonads may be cultured from many pieces of hospital equipment and not infrequently cultured from patients on whom such equipment is being utilized, such coincidental isolation does not establish a causal relationship since pseudomonads are ubiquitous in the hospital environment. Observations derived from data submitted from 69 hospitals in the USA which are reported to the Center for Disease Control as part of a National Nosocomial Infections Study indicate an overall rate for hospital-associated infections of 3·3–3·6 per 100 discharged patients in the last quarter of 1972 (Center for Disease Control, 1974). Within these overall figures, the rate for infections of the respiratory tract was 0·6–0·7 per 100 discharged patients. Forty-five per cent of the hospital-associated pneumonias were associated with aerobic Gram-negative bacilli. However, these data do not allow definition of the role of inhalation therapy since pneumonia caused by aerobic Gram-negative bacilli can occur in patients who have not been hospitalized or have not received ventilatory support. Consideration of the epidemiology and pathogenesis of hospital-associated pneumonia may assist in elucidation of the reasons for the present lack of precise data that define the risk.

## Pathogenesis of Hospital-Associated Bacterial Pneumonias

Bacteria may invade the alveolar level of the lung in sufficient density to produce pneumonia by one of three routes: by way of the vasculature, by aspiration from the pharynx, or suspended in inhaled gas. Invasion of the lung by aerobic Gram-negative bacilli through the pulmonary vasculature is possible when there is a peripheral site of infection causing bacteremia, such as *E. coli* pneumonia complicating pyelonephritis or *Ps. aeruginosa* pneumonia in the burned patient (Tillotson & Lerner, 1967; Teplitz, 1965). However, in the majority of patients with Gram-negative bacillary pneumonia no distant locus can be identified.

In circumstances where individuals are not receiving inhalation treatments, aerosol transmission is probably of little importance in the epidemiology of hospital-associated Gram-negative bacillary pneumonia. The bacterial density in air, even in hospitals, is quite low and most of the suspended bacteria are not Gram-negative bacilli (Green et al., 1962). Most bacterial pneumonias are due to micro-organisms that make up the flora of the pharynx. Indirect evidence supports this hypothesis. Pneumococci instilled into the nose of unanesthetized rabbits have been noted to appear in the lung within minutes (Cannon & Walsh, 1937). When bronchi of dogs were occluded with sterile cotton plugs and atelectasis was allowed to occur, pharyngeal organisms were recovered distal to the occlusion (Lansing & Jamieson, 1963). Culture of human lungs at autopsy has revealed bacterial species similar to those cultured in the pharynx (Smillie & Duerschner, 1947; Kneeland & Price, 1960; Knapp & Kent, 1968; Mays et al., 1969). It has been suggested that these findings are due to agonal or post-mortem invasion of the tracheobronchial tree (Norris & Pappenheimer, 1905). However, the findings are also compatible with the terminal failure of clearance mechanisms that had suppressed bacterial growth in the lungs during life. The mechanism by which bacteria spread from the pharynx to the lung during life remains unclear. The aspiration into the lung of radiopaque material instilled into the oropharynx of normal sleeping adults has been demonstrated (Amberson, 1937) but not confirmed (Winfield et al., 1973).

The oropharynx of a normal person apparently does not provide a suitable environment for the growth of aerobic Gram-negative bacilli; only about 2 per cent of normal persons harbour such organisms at any particular time (Johanson et al., 1969). When multiple cultures are performed on normal persons, the cumulative percentage of subjects with at least one positive culture increases, but previously positive persons usually are negative. This indicates that colonization is transient in normal persons, although the length of colonization has not been defined. Furthermore, massive exposure of normal persons to these organisms does not result in colonization of the upper respiratory tract (Bloomfield, 1920; Ostrom et al., 1958; Meyers et al., 1961). Thus, normal persons are only at slight risk (perhaps 2 per cent are at risk) of developing Gram-negative bacillary pneumonia.

Chronically or severely ill patients lose effective pharyngeal clearance mechanisms, allowing colonization with Gram-negative bacilli (Johanson et al., 1969). Exposure to such organisms from an exogenous source is evidently not necessary, because it has been found that approximately 20 per cent of patients sufficiently ill to be admitted to a medical intensive care unit were colonized at the time of admission (Johanson et al., 1972). Although previous therapy with antimicrobial drugs facilitates colonization and possibly pneumonia, it is, similarly, not a prerequisite (Philp & Spencer, 1974).

As anticipated from the concentration of chronically and severely ill patients in a hospital setting, Gram-negative bacillary oropharyngeal colonization is especially prevalent among hospitalized patients (Johanson et al., 1969, 1972). There is a rapid rise in the incidence of colonization for the first few hospital days, suggesting that there is a susceptible pool of patients who are particularly predisposed to colonization. The patients most liable to colonization are those with features suggesting severity of illness, such as coma, hypotension, acidosis, azotemia, and alterations in the leucocyte count, in whom the incidence of colonization approaches 75 per cent (Johanson et al., 1972). An increased prevalence of colonization among patients with respiratory disease or endotracheal intubation suggests that factors impairing lung clearance may also promote colonization. Patients who, although hospitalized, are not critically ill have a far lower prevalence of oropharyngeal colonization, approximately 30–40 per cent (Johanson et al., 1969). Although definitive data are not at hand, the occurrence of clusters of patients colonized with the same species of Gram-negative bacilli suggests that, in part, the bacteria are transmitted from patient to patient within the hospital setting. However, the 20 per cent incidence of colonization at the time of admission and the subsequent colonization of some patients with bacterial species not isolated from other patients in the same environment suggest that they are colonized with indigenous organisms. The gastro-intestinal tract is the most likely site from which the bacilli spread. Thus, even in the absence of contaminated ventilatory equipment, patients will be encountered with pneumonia caused by aerobic Gram-negative bacilli. In one recent observation on the occurrence of colonization of the oropharynx with aerobic Gram-negative bacilli, inhalation was not a factor which could be significantly correlated with colonization (Johanson et al., 1972).

Ventilatory equipment may serve as a reservoir from which patients become colonized with bacteria through several routes. Surface contamination of inhalation therapy equipment could serve as a potential source from which the patient's face, mouth or nose could become colonized; this has not been demonstrated directly. The exhalation side of the circuit has received little attention; however, effluent may contaminate the patient's immediate surroundings such as bed linens (Deane et al., 1970). From the contaminated local environment, secondary transmission by the patient's own hands or by nursing personnel to the index patient or other patients can occur (Babington et al., 1971). However, the most important mechanism for the dissemination of bacteria by ventilatory equipment is within aerosols which may pass into the stream of gas delivered by such equipment in particle sizes that are sufficiently small to be deposited in terminal lung units where host defence mechanisms may be overcome more easily. Studies on the infectivity of aerosols have shown that the minimum infectious dose of certain pathogenic bacteria is less if delivered in aerosols of a size capable of deposition beyond the level of ciliated epithelium (Sawyer, 1963). It has been estimated that at least 50 per cent of 1·0–2·0 $\mu$m particles, when delivered to the mouth or nose of an individual, are capable of entering the broncial tree distal to the terminal bronchioles (lowest level of ciliated epithelium) (Weimer et al., 1963; Mitchell, 1960). Most types of nebulizers are designed to deliver aerosols in this size range to assure humidification or delivery of medications to the lower reaches of the tracheobronchial tree. In our initial observations we demonstrated that the reservoir nebulizer was the major site which became contaminated and from which bacterial dissemination occurred (Reinarz et al., 1965). The conclusions were based on the observations derived from sampling the effluent gas by using an air sampler. The following observations were made. Equipment which did not incorporate reservoir nebulizers into its design did not generate significant bacterial aerosols. Second, sequential disassembly of apparatus with reservoir nebulizers demonstrated bacterial aerosols distal but not proximal to the reservoir nebulizer. Third, there was correlation of both numbers and species of organisms in the generated aerosols and the reservoir nebulizer fluid. The survival and multiplication of some bacterial species in the reservoir nebulizer fluid provides an effective amplification mechanism whereby even small numbers of initial contaminating bacteria ultimately may be delivered in large quantities to a patient. There are multiple potential routes by which the fluid-containing reservoir initially may become contaminated. The relative importance of one route over another cannot be generalized from available data. The sources to be considered include, at least, the following. The oxygen, compressed air or other gas which is administered to the patient and which often drives the nebulizer may be contaminated. Room air which usually is added through dilutor ports to decrease the concentration of inspired oxygen may provide another source. Personnel may contaminate the inside of the reservoir container during cleaning or replenishing the fluid in the reservoir. We even have had personnel contaminate the reservoir by attempting to be more cautious than called for in cleansing procedures and washing the jars with 3 per cent hexachlorophene-containing soaps which were unfortunately contaminated. The water or other fluid placed in the reservoir must be sterile. There are numerous outbreaks of pulmonary colonization and infection which have been traced to the nebulization of medications which were contaminated (Mertz et al., 1967; Cabrera, 1969; Sanders et al., 1970). Organisms present in secretions from the patient's nose or mouth may contaminate the fluid which tends to condense in the plastic delivery tubing connecting the nebulizer to the face mask.

Bacteria have been shown to survive for considerable time in such fluid. Massive contamination may result from emptying the condensate that collects in the delivery tubing back into the reservoir. Failure to disinfect portions of the nebulizer such as the nebulizer jet may occur because of entrapment of air which prevents contact with liquids if such are being used as disinfectants. Under these circumstances, the nebulizer jet serves as a nidus from which re-inoculation of the reservoir may occur.

Whichever primary mechanisms are operative, even reservoir nebulizers that are initially sterile and used by only one individual very frequently become contaminated within 24 hours of clinical use. The bacterial multiplication that then occurs results in an amplification of organisms that can be delivered in larger numbers and in a more infectious particle size (Reinarz et al., 1965; Edmondson et al., 1966a).

## Techniques for the Sampling of Inhalation Therapy Equipment

Another of the problems in an assessment of the hazard of infection associated with inhalation therapy equipment are the various techniques employed for bacteriologic sampling. Most studies have been based upon cultures obtained by means of swabbing or rinse techniques. While such techniques provide information, they do not demonstrate whether or not such organisms can be added by aerosol to the gases passing through the tubing and thus be inhaled by the patient. As noted by Magath, bacteria in water will not be transmitted unless the water is sprayed into the air (Lundy, 1943). The same seems to hold for bacteria on the surfaces of tubing, etc. Jenkins and Edgar employed a slit sampler to study the expiratory gases from an anesthetic apparatus into which the laboratory staff breathed (Jenkins & Edgar 1964). They compared the results obtained with the slit sampler with broth washings of the mask and tubings. With breathing, broth washings of the mask were positive from 10 of 11 subjects, while with the slit sampler only 1 of the 11 studies revealed organisms. Unfortunately, the researchers did not present any quantitative data. However, it is apparent from their data that while organisms may be present in the system as demonstrated by swabbing or rinsing techniques, they are infrequently recovered from effluent gases as revealed by quantitative air sampling techniques. Air sampling is best performed by a device that allows enumeration of the suspended bacteria and sizing the particles in which the bacteria are contained, such as the Andersen air sampler

(Andersen, 1958). Since such devices are rather cumbersome for routine clinical use, several simple sampling techniques have been devised which are sufficiently sensitive and accurate to be clinically most useful. We have developed and used a procedure in which a glass funnel and screw-top bacteriological tubes are used (Edmondson & Sanford, 1966). The usefulness of this procedure has been confirmed by Nelson and Ryan (Nelson & Ryan, 1971). Subsequently, an equally simple and more sensitive method has been published, although a comparison with standard air sampling techniques was not included (Nazemi et al., 1972).

While we are of the opinion that the use of air sampling techniques is the more appropriate means to assess the potential infectious hazard of inhalation therapy equipment, we recognize that if cultures of reservoir liquids and tubing surfaces are free of bacteria, this provides excellent evidence that the effluent gas will not be contaminated. The problem arises over the interpretation of positive cultures, which may or may not be indicative of a problem and because of their frequency may result in a sense of futility with abandonment of a disinfection programme.

## Basic Designs of Inhalation Therapy Equipment

Since gases such as oxygen or compressed air are essentially free of water vapour, administration without humidification results in desiccation of the upper airways and may significantly predispose to infection. Hence, a method for increasing the water vapour content of administered gases is essential even though it results in the concomitant increased risk of bacterial contamination of the moisture source. There are two markedly differing principles utilized to increase the humidity of gases delivered to the tracheobronchial tree. The first of these involves humidification, the second nebulization. Definition of these terms becomes essential since differences in terminology with resultant confusion exists in the literature.

Humidifiers are devices that saturate gas with water vapour; they do not aerosolize droplet water. Most clinical humidifiers cause the stream of gas to bubble through water, although some humidifiers used in paediatrics cause the stream of gas to be blown across the surface of the water.

In contrast, nebulizers not only saturate gas with water vapour, but also disperse an aerosol of droplets. Most clinical nebulizers are gas driven, operating on the Venturi principle. Venturi nebulizers may be of small volume, 5–30 ml for dispensing specific medication or they may contain a large reservoir, approximately 500 ml, for the administration of moisture. These latter nebulizers may be operated at room temperature or they may be heated. Aerosols also may be produced by dropping liquid onto the surface of a rapidly spinning disc; such nebulizers, which are infrequently used in adult medicine, also incorporate large-volume reservoirs (Grieble et al., 1970). A third means of creating aerosols is by ultrasonic nebulizers, in which droplets are produced by a rapidly oscillating crystal (Ringrose et al., 1968).

Disinfection of each type of equipment and even different devices employing the same design principle must be evaluated independently.

## Microbiological Observations

The first systematic investigations of decontamination of effluent gas from inhalation therapy equipment were reported in 1965 and 1966 (Reinarz et al., 1965; Edmondson et al., 1966a). It was found that the effluent gas from low flow humidifying devices (flow 15 l/min or less) contained no more bacteria than did hospital air. These observations have been extended to humidifiers capable of saturating gases with water vapour at a very high flow and, hence, capable of use with intermittent positive pressure breathing (IPPB) machines (Schulze et al., 1967). These high flow humidifiers generated greater bacterial counts than room air only 9 per cent of the time, and this occurred only when the liquid in the humidifier was massively contaminated to $10^6$ or greater organisms per millilitre. The relative inability of humidifiers to aerosolize bacteria is probably related to the lack of liquid droplets. These results indicate that humidifiers in clinical practice need only reasonable care. When the water in the reservoir jar has fallen to a low level, it should be discarded and replaced with fresh water; the humidifiers should be sterilized by appropriate means when changed from one patient to another. However, more rigorous daily decontamination or sterilization of such humidification equipment does not seem essential.

Gas driven, small-volume (medication) Venturi nebulizers used with IPPB machines are capable of generating bacterial aerosols, but they were found to do so infrequently in hospital practice (Reinarz et al., 1965), and subsequent experience has confirmed these observations (Pierce et al., 1970). We assume that this is true because the medication introduced into the nebulizer is sterile, and the small volume of the nebulizer allows it to dry between treatments. There have been reported outbreaks of hospital-acquired infection due to medication nebulizers (Mertz et al., 1967; Cabrera, 1969; Sanders et al., 1970). In each instance the source of the bacteria was found to be contaminated medications placed into the nebulizers. At present, it is reasonable to believe that medication nebulizers are free from significant danger, provided that the medications to be used in them are handled in an aseptic manner, the nebulizers are dried between treatments, and sterilized by appropriate means between patients.

Reservoir Venturi nebulizers, operating either at ambient or body temperature, have been demonstrated to be a major potential source of bacteria contained in particle sizes capable of penetrating terminal lung units (Reinarz et al., 1965; Edmondson et al., 1966a; Moffet & Williams, 1967; Moffet et al., 1967). Bacterial contamination occurs despite soaking the nebulizers in disinfectants and recurs promptly when the nebulizers are put into clinical use even when they are sterile at the onset of treatment (Reinarz et al., 1965; Edmondson et al., 1966a). Contamination occurs regardless of the therapeutic solution used and irrespective of whether the nebulizers are used alone or with IPPB devices. Such aerosol contamination has been demonstrated to be an important cause of patient morbidity and mortality (Pierce et al., 1970; Phillips & Spencer, 1965; Pierce et al., 1966; Moffet & Allan, 1967; Phillips, 1967; McNamara et al., 1967; Cabrera, 1969).

Spinning disc nebulizers (Grieble et al., 1970) and ultrasonic nebulizers (Ringrose et al., 1968; Rhoades et al., 1971) also have been demonstrated to produce bacterial aerosols, and such aerosols have been incriminated in the colonization and infection of patients with Gram-negative bacilli. These types of nebulizers also contain large reservoirs of liquid to be nebulized.

The bacteria that are found are virtually always aerobic Gram-negative bacilli. The bacterial species isolated in our initial studies are tabulated in Table 3.1 (Reinarz et al., 1965). Multiple species of organisms were isolated more commonly than a single species. In no samples were Gram-positive organisms recovered. This general pattern has held in all subsequent studies which are based upon aerosol samples, although Nelson and Ryan isolated Gram-positive organisms from aerosol samples from 8 of 27 patients: *Staph. epidermidis* 5 isolates. *Staph. aureus* 3 isolates. viridans streptococci 2 isolates. hemolytic streptococci 1 isolate.

Because of the absence of Gram-positive cocci in our initial studies and since several patients had staphylococcal pneumonia at the time inhalation therapy was instituted. we studied the persistence of *Staph. aureus*. strain 502A. following massive contamination of a reservoir

**Table 3.1.** *Results of Bacterial Cultures of Inhalation Therapy Equipment at Various Hospitals*

| Hospital | Source of cultures | Number of positive cultures | Bacterial species (number of isolates) | | | | | | | | | Prevalent organism |
|---|---|---|---|---|---|---|---|---|---|---|---|---|
| | | | Pseudomonas spp. | Flavobacterium meningosepticum | Acinetobacter calcoaceticus var. anitratus | Alcaligenes fecalis | Achromobacter spp. | Acinetobacter calcoaceticus var. lwoffi | Serratia marcescens | Enterobacter aerogenes | | |
| A | Aerosol | 62 | 32 | 12 | 20 | 27 | 15 | 3 | 3 | 4 | | Pseudomonas aeruginosa |
| | Nebulizer fluid | 44 | 21 | 7 | 5 | 22 | 5 | 3 | 3 | 4 | | |
| B | Aerosol | 20 | 4 | 15 | 2 | 9 | 4 | 1 | 0 | 0 | | Flavobacterium meningosepticum |
| | Nebulizer fluid | 22 | 9 | 12 | 0 | 5 | 1 | 2 | 0 | 0 | | |
| C | Aerosol | 6 | 3 | 2 | 0 | 1 | 1 | 0 | 0 | 0 | | Pseudomonas aeruginosa |
| | Nebulizer fluid | 7 | 4 | 2 | 0 | 2 | 1 | 1 | 0 | 0 | | |
| D | Aerosol | 13 | 3 | 2 | 11 | 1 | 2 | 3 | 0 | 0 | | Acinetobacter calcoaceticus var. anitratus (Herellea) |
| | Nebulizer fluid | 9 | 1 | 2 | 8 | 1 | 1 | 0 | 0 | 0 | | |
| E | Aerosol | 20 | 18* | 5 | 0 | 2 | 0 | 1 | 0 | 0 | | Pseudomonas species* |
| | Nebulizer fluid | 19 | 18* | 1 | 0 | 0 | 0 | 0 | 0 | 0 | | |

Isolates studied by Dr S. G. Carey at Walter Reed Army Institute of Research were considered to closely resemble *Cytophaga johnsonae*.

* Isolates studied at Center for Disease Control and reported as Pseudomonas species.

nebulizer. The reservoir nebulizer jars of three IPPB machines were filled with cultures of viable *Staph. aureus* 502A. The number of organisms varied from $1.3 \times 10^9$ to $2.5 \times 10^9$ viable units per ml. After flushing with sterile distilled water, viable staphylococci in the aerosols exceeded 10000 colony forming units per cubic foot of air. After 8h. species of aerobic Gram-negative bacilli were isolated from the aerosols generated by each of the machines. Furthermore, the number of staphylococci was less than 500 colony forming units per cubic foot. By 24 h. staphylococci were no longer isolated but, in contrast, large numbers of aerobic Gram-negative bacilli (*Acinetobacter calcoaceticus var. anitratus* and *Achromobacter* spp) were isolated. This lack of persistence of staphylococci should not have been unanticipated as McDade and Hall have shown that the survival of strains of *Staph. aureus* was diminished at high relative humidity (McDade & Hall. 1963).

The ability of various micro-organisms to proliferate in the fluid in the reservoir nebulizer is important in determining the organisms that predominate. Maki and Martin (1975) have shown that organisms of the tribe Klebsielleae (*Klebsiella, Enterobacter,* and *Serratia*) can multiply in 5 per cent dextrose in water while pseudomonads do not. Favero and associates found that naturally occurring strains of pseudomonads. i.e., strains inoculated directly from water sources in the hospital, multiply rapidly in distilled water (Favero et al.. 1971). Interestingly, strains precultured in broth before inoculation grew poorly. While we observed that the strains of *Serratia marcescens.* serotype 014 : H4. isolated in an outbreak of infection due to contaminated inhalation therapy medications, especially Alevaire (Tyloxapol). would multiply in this product, intensive study of the growth characteristics of the many bacterial species associated with inhalation therapy has not been reported (Sanders et al.. 1970).

## Disinfection of Inhalation Therapy Equipment

Several methods have been suggested for sterilization of inhalation therapy equipment. Based on surface culturing techniques, the mechanical ventilator. rather than the reservoir nebulizer, is frequently the focus of major concern (Bishop et al.. 1962; Bishop et al.. 1966a; Sykes, 1964; Gullero et al., 1967; Spencer et al., 1968; Judd et al., 1968). Although ventilators. like other hospital equipment, should receive reasonable hygienic care, they are not a major source of bacterial contamination of inspired gas (Reinarz et al.. 1965). The use of bacterial filters in various positions of the stream of gas flow in ventilators also has been suggested (Bishop et al., 1963; Bishop, 1966; Hellewell et al., 1967). A filter on the driving gas of the ventilator or nebulizer eliminates only one source of potential contamination, and, hence, filters do not ensure sterility of the effluent gas from the nebulizer. Filters interposed between the nebulizer and the patient are not possible, since they do not permit the passage of the therapeutic aerosol. The primary attention must be directed to the decontamination of the reservoir nebulizer.

In the published material on inhalation therapy equipment, varied terminology has been utilized to describe the processes utilized to prevent transmission of disease. We have used the term decontamination for many of the procedures; however, according to common usage this is inappropriate and the term decontamination should be reserved for the process that renders contaminated items safe to handle with reasonable care (Mallison, 1974). Disinfection, the more correct term, may be defined as the virtual elimination of all harmful micro-organisms, except spores, in an attempt to prevent transmission of disease. Sterilization is the complete destruction or removal of all organisms. Objects that contact skin or mucous membranes such as respiratory therapy equipment other than the apparatus itself should receive at least a high level of disinfection and ideally should be sterilized.

There are three methods potentially available for the sterilization of inhalation therapy equipment, steam autoclaving, cold soaking with activated glutaraldehyde germicidal solution, and ethylene oxide gas (Engley, 1970). Steam autoclaving is usually not practical because of the sensitivity of the equipment to heat, although we have utilized this procedure successfully for older types of nebulizers which were constructed of cast aluminium (Edmondson et al., 1966a).

Immersion in a germicidal solution may be used to either disinfect or sterilize equipment, but 10 h submersion in activated glutaraldehyde is required for sterilization; less time results in disinfection (Engley, 1970). The greatest limitation to liquid sterilization (or disinfection) is its unreliable penetration. To sterilize an item the active chemical must contact all bacteria-laden surfaces. Entrapment of air often prevents adequate liquid contact during treatment and is a particular problem with inhalation therapy equipment as nebulizer jets are usually encased in hollow areas.

Since gases are completely miscible with air, the ideal chemical sterilant is a gas, such as ethylene oxide. Ethylene oxide denatures bacterial proteins through a process of alkylation. Moisture is essential to this reaction. Raising the temperature also hastens the reaction, sterilization being achieved 6–8 times faster at 54·5°C (130°F) than at

21°C (70° F) (Engley, 1970). Ethylene oxide is absorbed into various plastics and must be allowed to diffuse out before the plastic is in prolonged contact with tissues to avoid its local irritant effect; this process is termed aeration. Aeration periods depend on several factors: duration of exposure to ethylene oxide, the aeration environment, type of plastic and intended use of the item. While aeration may often require 24–48 hours, units are available to reduce all aeration periods to eight hours. The major problem of ethylene oxide sterilization is the expense of the equipment required to assure safety and prompt effective steriliza- ation and aeration. However, this technique appears to be the most satisfactory provided it is remembered that under in-use conditions recontamination can occur within 24 h even if a piece of equipment is used by only one patient. Lack of appreciation of this is illustrated by the suggestion that the problem of pseudomonas infection could be avoided by the use of aerosol equipment kept in the patient's room, rather than by the use of IPPB machines which are shared by several patients (Skinner, 1968).

Since many hospitals still do not have the capability to sterilize all their inhalation therapy equipment with ethylene oxide and it is not feasible to aerosolize the most effective liquid chemical sterilant, activated glutaraldehyde, to assure that the nebulizer jets are sterilized, various disinfecting procedures have been employed and have been shown to be effective if rigorously applied and monitored. With these procedures, it is important to mechanically cleanse the equipment. Exudates should not be allowed to dry. Equipment should be rinsed initially with cold water to avoid coagulation of protein, then washed with mechanical cleansing and detergent. Heating to 75°C (167°F) for 10 min will kill vegetative forms of bacteria and even washing at 60°C (140°F) for 2 min destroys most bacteria (Jenkins & Edgar, 1964; MacCallum & Noble, 1960). The use of hot water, 'pasteurization', has been successfully employed by several groups (Roberts et al., 1969; Nelson & Ryan, 1971). Roberts and associates immersed equipment for a period of 15 min at 80–85°C (176–185°F) while Nelson and Ryan processed equipment for 30 min at 70°C (158°F). Nelson and Ryan were concerned that jets in equipment might not be disinfected by this method as it was questionable whether the water would enter the tiny orifices. All items with small orifices or jets were disassembled and disin- fected with activated glutaraldehyde (immersed for 10 min), rinsed in a solution of one ounce of sodium bisulphite per gallon of water, then pasteurized and air-dried by blowing 100°F (37·8°C) air over the equip- ment.

Alternative methods of disinfecting equipment have been utilized.

We have employed the nebulization of 0·25 per cent acetic acid as an adjunct disinfecting procedure coupled to the use of ethylene oxide, glutaraldehyde or soaking in a phenolic disinfectant solution. Other procedures include the use of chlorhexidine, higher concentrations of acetic acid, 7·5 per cent hydrogen peroxide, copper sponges, or dilute silver nitrate.

The most extensively reported means of sterilizing reservoir Venturi nebulizers involves a programme which consists of sterilization of the equipment with ethylene oxide or steam autoclaving or immersion in a disinfectant solution such as glutaraldehyde or a phenolic disinfectant in the inhalation therapy department combined with the daily nebulization of 0·25 per cent acetic acid when the equipment has been assigned to a patient (Reinarz et al., 1965; Edmondson et al., 1966a; Pierce et al., 1970). The details of this regimen are as follows:

(1). If the nebulizer is attached to the patient, turn off, and remove the face mask (or mouthpiece) from the patient.

(2). Remove the nebulizer jar and empty the contents.

(3). Rinse the jar with sterile water.

(4). Cover the fluid intake of the tube leading to the nebulizer jet with 0·25 per cent acetic acid (approximately 200 ml in most nebulizers).

(5). Replace the jar on the nebulizer.

(6). Turn on the nebulizer to a flow rate of 7 l/min, so that the acetic acid is aerosolized. This can be nebulized in the room with the patient disconnected from the tubing without causing unpleasant odours or irritation of mucous membranes.

(7). Allow the acetic acid to nebulize for approximately 10 min.

(8). Turn off the gas (oxygen or air) and remove the nebulizer jar.

(9). Discard the acetic acid, rinse the jar with sterile water.

(10). Refill the jar with sterile water or other desired solution. Be careful not to contaminate the inside of the jar in the process.

(11). The equipment is now ready for patient use.

(12). The nebulizer fluid should be emptied and replenished every eight hours. At this time always discard the remainder from the jar to minimize any residual inoculum.

(13). Before the equipment is used on another patient, it should be sterilized or disinfected. If the use by one patient is prolonged, equipment should be returned and newly sterilized equipment issued every 72 h.

Dilute acetic acid was chosen since it is relatively nontoxic, enabling the procedure to be performed at the patient's bedside. Parker and Hoeprich (1963) found that acetic acid was bactericidal for only 49 of 149 strains of Gram-negative bacilli and Gram-positive cocci. In vitro studies in our laboratory have confirmed these observations. Neverthe-less, microbiological failure of the system of decontamination has not been encountered. When properly used, the regimen of acetic acid renders the effluent gas from most Venturi nebulizers virtually sterile. It is perhaps of note, however, that when the acetic acid method was introduced from the laboratory into routine hospital use, the results were disappointing (Pierce et al., 1970). It was found subsequently that the failure to achieve decontamination of effluent gas from nebulizers resulted from human error or faulty nebulizer jets. A continuous monitor-ing programme, and replacement of pitted nebulizer jets, resulted in the equipment being contaminated less than 10 per cent of the time while in use. When treatment with a Venturi nebulizer is discontinued in a given patient, it is sterilized by autoclaving, ethylene oxide, or by soaking in a disinfectant followed by aerosolization with acetic acid.

It is not possible to decontaminate all types of reservoir Venturi nebulizers by the daily acetic acid regimen (Edmondson et al., 1966b). Some nebulizers used in pediatric tents have been demonstrated to be resistant to this method of decontamination. It has been found that such nebulizers must be sterilized by some other means at least every 48 h; either autoclaving or ethylene oxide may be used with acetic acid nebulization on alternate days.

Limited studies with the nebulization of other disinfectant solutions have been reported (Phillips & Spencer, 1965; Edmondson et al., 1966b). As noted by Parker and Hoeprich (1963), in vitro chlorhexidine diacetate appeared to be more effective than acetic acid. However, the potential toxicities of chlorhexidine and similar agents following nebulization into areas, or inadvertently into patients, have not been sufficiently defined to allow their usage without considerable study.

It has been reported that the placing of a copper sponge in the reservoir jars of Venturi nebulizers results in the nebulizer contents remaining sterile (Deane et al., 1970). The institution of this method of cleaning into an intensive care unit resulted in an apparent reduction in the incidence of pulmonary infections. Although these investigators concluded that sterility of reservoir nebulizers could be maintained for 48 h or longer, they note that in practice the nebulizers and all parts distal to them are changed every 24 h, a procedure that should result in sterile equipment. Our own experience with this method has been limited to intentional bacterial inoculation of nebulizer contents with observations on the ability of the copper sponge to sterilize the solutions. We have been disappointed by the erratic ability of copper sponges to maintain sterility.

Grieble et al., (1970) have found that daily use of phenolic disinfectants or dilute acetic acid did not sterilize spinning disc nebulizers. It was found that daily sterilization with ethylene oxide was necessary to prevent contamination. Since this procedure was considered impractical in their clinical setting, the use of spinning disc nebulizers was discontinued in their hospital.

Rhoades et al., (1971) have found that ultrasonic nebulizers were not sterilized by soaking in a disinfectant solution or by the nebulization of 0·25 per cent acetic acid for 30 min. However, 2 per cent acetic acid nebulization for 30 min was successful in decontaminating a nebulizer. It has also been found that 7·5 per cent hydrogen peroxide nebulization for 20 min may be an effective decontamination procedure for ultrasonic nebulizers (Stevens et al., 1970). It is suggested that such equipment be decontaminated at least once every 24 h (Rhoades et al., 1971).

## Results of Programmes of Disinfection of Inhalation Therapy Equipment

During the period immediately preceding the institution of a programme of culturing the effluent from inhalation therapy equipment and disinfection employing the procedure which involves acetic acid nebulization, the prevalence of necrotizing pneumonia was 7·9 per cent (Pierce et al., 1970). Under the epidemiologic circumstances of the hospitalized patient, the histologic appearance of necrotizing pneumonia has been found to be specific enough to serve as an index of the frequency of all pneumonia due to aerobic Gram-negative bacilli (Mays et al., 1969). The prevalence of this lesion increased from 1·8 per cent of all autopsies in 1957 to 7·9 per cent of autopsies in 1963 (Pierce et al, 1966). This increase was significantly correlated with the use of reservoir nebulization treatments. With a programme of surveillance and disinfection, and in the face of a 9 per cent increase in the use of reservoir nebulization treatments, the incidence of necrotizing pneumonia in 1966 and 1967 decreased to 2·2 and 2·1 per cent of autopsies. This accompanied a decrease in the frequency of contamination of reservoir nebulizers from 84 per cent to less than 10 per cent. Subsequently, in a totally independent study of the occurrence of colonization of the oropharynx and the development of pneumonia in patients in a medical intensive care unit, inhalation treatments were not significantly associated with the acquisition of aerobic Gram-negative bacilli.

The information available justifies several conclusions. Inhalation therapy equipment is a potential vector of Gram-negative bacilli. However, not all types of equipment are equally dangerous in this regard, and investigators should be more specific in their definitions of equipment used when reporting results. Ventilators and humidifiers have minimal potential for aerosolizing bacteria into the patient's inspired gas. Small-volume Venturi nebulizers are potential disseminators of bacteria, but they evidently infrequently do so in hospital use. Large-volume nebulizers of all designs (Venturi, spinning disc, or ultrasonic) are the major source of aerosolized bacteria. The means required for sterilization differ depending on the design of the nebulizer. It has been demonstrated that most nebulizers can be decontaminated successfully in a clinical setting. However, any system of decontamination requires meticulous handling by personnel and the application of some type of decontamination procedure at least once ever 24 h. No hospital can consider its own procedure satisfactory unless it has a continuous monitoring programme to demonstrate that the degree of contamination does not exceed that of hospital air. These observations indicate the need for the same surveillance of this form of therapy as is carried out in many other facets of patient care.

## Acknowledgments

The studies upon which this presentation is primarily based represent the concerted efforts of many individuals with whom we have had the privilege of working. We wish to recognize their invaluable contributions: J. A. Reinarz, M.D., E. B. Edmondson, M.D., Benita, B. Mays, M.A., Grace D. Norris, R.N., Carolyn G. Wickwire, B.A., P. M. Southern, Jr., M.D., C. V. Sanders, Jr., M.D., W. G. Johanson, Jr., M.D., T. Schulze, M.D., J. S. Leonard, Jr., M.D., G. McGee, M.D., and J. Ketchersid, M.D.

The authors' studies were supported by Public Health Service Research Grant CC 00202 from the Center for Disease Control, Atlanta, Georgia, and grants T01 A1 00030 from the National Institute of Allergy and Infectious Diseases, and HL 14187 (SCOR) from the National Heart and Lung Institute, National Institutes of Health, Bethesda, Maryland, USA.

## References

Adriani, J., & Rovenstine, E. A. (1941) Experimental studies on carbon dioxide absorbers for anesthesia. *Anesthesiology*, **2**, 1.
Amberson, J. B., Jr (1937) Aspiration bronchopneumonia. *Int. Anesthesiol. Clin.*, **3**, 126.

Andersen, A. A. (1958) New sampler for the collection, sizing and enumeration of viable airborne particles. *J. Bact.,* **76**, 471.

Babington, P. C. B., Baker, A. B. & Johnston, H. H. (1971) Retrograde spread of organisms from ventilator to patient via the expiratory limb. *Lancet,* **1**, 61.

Bishop, C., Potts, M. W. & Molloy, P. J. (1962) A method of sterilization of the Barnet respirator. *Br. J. Anaesth.,* **34**, 121.

Bishop, C., Roper, W. A. G. & Williams, S. R. (1963) The use of an absolute filter to sterilize inspiratory air during intermittent positive pressure respiration. *Br. J. Anaesth.,* **35**, 32.

Bishop, C. (1966) *Pseudomonas aeruginosa* cross-infection. *Lancet,* **1**, 267.

Bishop, C., Robertson, D. S. & Williams, S. R. (1966) The use of ethylene oxide for sterilization of mechanical ventilators. *Br. J. Anaesth.,* **36**, 53.

Bloomfield, A. L. (1920) The fate of bacteria introduced into the upper air passages. V. The Friedlander bacilli. *Bull. Johns Hopkins Hosp.* **31**, 203.

Cabrera, H. A. 1969. An outbreak of *Serratia marcescens* and its control. *Arch. Int. Med.,* **123**, 650.

Cannon, P. R. & Walsh, T. E. (1937) Studies on the fate of living bacteria introduced into the upper respiratory tract of normal and intranasally vaccinated rabbits. *J. Immun.,* **32**, 49.

Center for Disease Control, (1974) *National Nosocomial Infections Study Quarterly Report, Fourth Quarter* 1972, Atlanta, Georgia.

Deane, R. S., Mills, E. L. & Hamel, A. J. (1970) Antibacterial action of copper in respiratory therapy apparatus. *Chest* **58**, 373.

Duncom, B. M. (1947) *The Development of Inhalation Anesthesia.* London: Oxford University Press.

Edmondson, E. B., Pierce, A. K. & Sanford, J. P. (1966a) *Pseudomonas aeruginosa* cross-infection. *Lancet,* **1**, 660.

Edmondson, E. B. & Sanford, J. P. (1966) Simple methods of bacteriologic sampling of nebulization equipment. *Am. Rev. resp. Dis.* **81**, 450.

Edmondson, E. B., Reinarz, J. A., Pierce, A. K. & Sanford, J. P. (1966b) Nebulization equipment. A potential source of infection in gram-negative pneumonias. *Am. J. Dis. Child.,* **111**, 357.

Engley, F. B. (1970) Ethylene oxide sterilization. A necessity for inhalation therapy departments. *Inhalation Ther.,* **15**, 9.

Favero, M. S., Carson, L. A., Bond, W. W. & Peterson, N. J. (1971) *Pseudomonas aeruginosa:* growth in distilled water for hospitals. *Science, N.Y.* **173**, 836

Green, V. M., Vesley, D., Bond, R. G. & Michaelsen, G. S. (1962) Microbiological contamination of hospital air. I. Quantitative studies. *Appl. Microbiol.* **10**, 561.

Grieble, H. G., Colton, F. R., Bird, T. J., Toigo, A. & Griffith, L. (1970) Fine-particle humidifiers. Source of *Pseudomonas aeruginosa* infections in a respiratory care unit. *New Engl. J. Med.* **282**, 531.

Gullers, K., Malmorg, A. S., Norlander, O., Nyström, B. & Peterson, N. (1967) Pseudomas infection in hospital. *Br. med. J.,* **4**, 548.

Hellewell, J., Jeanes, A. L., Walking, R. R. & Gibbs, F. J. (1967) The Williams bacterial filter, use in the intensive care unit. *Anaestheia,* **22**, 497.

Hoffman, M. A. & Finberg, L. (1955) Pseudomonas infections in infants associated with high humidity environment. *J. Pediatrics* **46**, 626.

Jenkins, J. R. E. & Edgar, W. M. (1964) Sterilization of anesthetic equipment. *Anaesthesia,* **19**, 177.

Johanson, W. G., Pierce, A. K. & Sanford, J. P. (1969) Changing pharyngeal bacterial flora of hospitalized patients. Emergence of gram-negative bacilli. *New Engl. J. Med.* **281**, 1137.

Johanson, W. G., Jr., Pierce, A. K., Sanford, J. P. & Thomas, G. D. (1972) Nosocomial respiratory infections with gram-negative bacilli. *Ann. Intern. Med.* **77**, 701.

Judd, P. A., Tomlin, A. J., Whitby, J. L., Englis, T. C. M. & Robinson, J. S. (1968) Disinfection of ventilators by ultrasonic nebulization. *Lancet*, **2**, 1019.

Knapp, B. E. & Kent, T. H. (1968) Post-mortem lung cultures. *Archs. Path.*, **85**, 200.

Kneeland, Y., Jr. & Price, K. M. (1960) Antibiotics and terminal pneumonia: A post-mortem microbiological study. *Am. J. Med.*, **29**, 967.

Lansing, A. M. & Jamieson, W. G. (1963) Mechanisms of fever in pulmonary atelectasis. *Archs. Surg.*, **87**, 168.

Livingstone, H., Heidrick, F., Holicky, I. & Dack, G. M. (1941) Cross infections from anesthetic face masks. *Surgery*, **9**, 433.

Lundy, J. S. (1943) *Clinical Anaesthesia.* Philadelphia: W. B. Saunders Co.

MacCallum, F. O. & Noble, W. C. (1960) Disinfection of anaesthetic face masks. *Anaesthesia* **15**, 307.

Macpherson, R. (1958) Oxygen therapy: an unsuspected source of hospital infections. *J. Am. med. Ass.* **167**, 1083.

Maki, D. G. & Martin, W. T. (1975) Nationwide epidemic of septicemia caused by contaminated infusion products. IV. Growth of microbial pathogens in fluids for intravenous infusion. *J. infect. Dis.*, **131**, 267.

Mallison, G. F. (1974) *APIC Newsletter* v. 2, #2.

Mays, B. B., Thomas, G. D., Leonard, J. S., Jr., Southern, P. M., Pierce, A. K. & Sanford, J. P. (1969) Gram-negative bacillary necrotizing pneumonia: a bacteriologic and histopathologic correlation. *J. infect. Dis.*, **120**, 687.

McDade, J. J. & Hall, L. B. (1963) Survival of *Staphylococcus aureus* in the environment. Exposure on surfaces. *Am. J. Hyg.*, **78**, 330.

McNamara, M. J., Hill, M. C., Barrows, A. & Tucker, E. B. (1967) A study of the bacteriologic patterns of hospital infections. *Ann. intern. Med.* **66**, 480.

Mertz, J. J., Scharer, L. & McClement, J. H. (1967) A hospital outbreak of klebsiella pneumonia from inhalation therapy with contaminated aerosol solutions. *Am. Rev. resp. Dis.*, **95**, 454.

Meyers, C. E., James, H. A. & Zippin, C. (1961) The recovery of aerosolized bacteria from humans. I. Effects of varying exposure, sampling times, and subject variability. *Arch. envir. Hlth*, **2**, 384.

Mitchell, R. I. (1960) Detention of aerosol particles in the respiratory tract. *Am. Rev. resp. Dis.*, **82**, 627.

Moffet, H. L. & Allan, D. (1967) Colonization of infants exposed to bacterially contaminated mists. *Am. J. Dis. Child.*, **114**, 21.

Moffet, H. L., Allan, D. & Williams, T. (1967) Survival and dissemination of bacteria in nebulizers and incubators. *Am. J. Dis. Child.*, **114**, 13.

Moffet, H. L. & Williams, T. (1967) Bacteria recovered from distilled water and inhalation therapy equipment. *Am. J. Dis. Child.*, **114**, 7.

Nazemi, M. M., Musher, D. M. & Martin, R. R. (1972) A practical method for monitoring bacterial contamination of inhalation therapy machines. *Am. Rev. resp. Dis.* **106**, 920.

Nelson, E. J. & Ryan, K. J. (1971) A new use for pasteurization: disinfection of inhalation therapy equipment. *Respiratory Care* **16**, 97.

Norris, C. & Pappenheimer, A. M. (1905) A study of pneumococci and allied organisms in human mouths and lungs after death. *J. Exp. Med.,* **7**, 450.

Ostrom, C. A., Wolochow, H. & James, H. A. (1958) Studies on the experimental epidemiology of respiratory disease. IX. Recovery of airborne bacteria from the oral cavity of humans: The effect of dosage on recovery. *J. infect. Dis.,* **102**, 251.

Parker, R. W. & Hoeprich, P. D. (1963) In vitro effect of buffered solutions of acetic acid, triclobisonium chloride, chlorhexidine diacetate, and chlorhexidine digluconate on urinary tract infections. *In*: Antimicrobial Agents & Chemotherapy—1962, p. 26. J. C. Sylvester, Ann Arbor: Braun-Brumfield.

Phillips, I. & Spencer, C. (1965) *Pseudomonas aeruginosa* cross-infections due to contaminated respiratory apparatus. *Lancet*, **2**, 1325.

Phillips, I. (1967) *Pseudomonas aeruginosa* respiratory tract infections in patients receiving mechanical ventilation. *J. Hyg.,* **65**, 229.

Philp, J. R. & Spencer, R. C. (1974) Secondary respiratory infection in hospital patients: effect of antimicrobial agents and environment. *Br. med. J.,* **2**, 359.

Pierce, A. K., Edmondson, E. B., McGee, G., Ketchersid, J., Loudon, R. G. & Sanford, J. P. (1966) An analysis of factors predisposing to gram-negative bacillary necrotizing pneumonia. *Am. Rev. resp. Dis.* **94**, 309.

Pierce, A. K., Sanford, J. P., Thomas, G. D. & Leonard, J. S. (1970) Long-term evaluation of decontamination of inhalation-therapy equipment and the occurrence of necrotizing pneumonia. *New Engl. J. Med.,* **282**, 528.

Reinarz, J. A., Pierce, A. K., Mays, B. B. & Sanford, J. P. (1965) The potential role of inhalation therapy equipment in nosocomial pulmonary infection. *J. clin. Invest.,* **44**, 831.

Rhoades, E., Ringrose, R., Mohr, J. A., Brooks, L., McKown, B. A. & Felton, F. (1971) Contamination of ultrasonic nebulization equipment with gram-negative bacteria. *Archs. intern. Med.,* **127**, 228.

Ringrose, R. E., McKown, B., Felton, F. G., Barclay, B. O., Muchmore, H. G. & Rhoades, E. R. (1968) A hospital outbreak of *Serratia marcescens* associated with ultrasonic nebulizers. *Ann. intern. Med.,* **69**, 719.

Roberts, F. J., Cockcroft, W. H. & Johnson, H. E. (1969) A hot water disinfection method for inhalation therapy equipment. *Can. Med. Ass. J.* **101**, 30.

Sanders, C. V., Luby, J. P., Johanson, W. G., Barnett, J. A. & Sanford, J. P. (1970) *Serratia marcescens* infections from inhalation therapy medications: nosocomial outbreak. *Ann. intern. Med.,* **73**, 15.

Sawyer, W. D. (1963) Airborne infection. *Milit. Med.* **128**, 90.

Schulze, T. E., Edmonton, E. B., Pierce, A. K. & Sanford, J. P. (1967) Studies of a new humidifying device as a potential source of bacterial aerosols. *Am. Rev. resp. Dis.,* **96**, 517.

Sever, J. L. (1959) Possible role of humidifying equipment in spread of infections from the newborn. *Pediatrics*, **24**, 50.

Skinner, E. F. (1968) Aerosol inhalations or IPPB? (Letter to the Editor) *New Engl. J. Med.,* **278**, 1403.

Smillie, W. G. & Duerschner, D. R. (1947) The epidemiology of terminal bronchopneumonia. II. The selectivity of nasopharyngeal bacteria in invasion of the lungs. *Am. J. Hyg.,* **45**, 13.

Spencer, G., Ridley, M., Eykyn, S. & Achong. J. (1968) Disinfection of lung ventilators by alcohol aerosol. *Lancet*, **2**, 667.

Stevens, H. R., Martin, R. A. & Adiska, T. R. (1970) Disinfection of inhalation therapy equipment by ultrasonic nebulization. *Inhalation Ther.* **15**, 29.

Sykes, M. K. (1964) Sterilizing mechanical ventilators. *Br. Med. J.*, **1**, 561.

Teplitz, C. (1965) Pathogenesis of pseudomonas vasculitis and septic lesions. *Archs. Path.*, **80**, 297.

Tillotson, J. R. & Lerner, A. M. (1967) Characteristics of pneumonias caused by *Escherichia coli. New Engl. J. Med.*, **277**, 115.

Weimer, J. T., Ballard, T. A. & Punte, C. L. (1963) Respiratory retention of one micron particles in man. *Dis. Chest*, **44**, 268.

Winfield, J. B., Sande, M. A. & Gwaltney, J. M., Jr. (1973) Aspiration during sleep. *J. Am. med. Ass.*, **223**, 1288.

# 4 | Advances in Bacterial Taxonomy

## L. R. HILL

## Introduction

This chapter is divided into three parts which follow the divisions of taxonomy itself: classification, nomenclature and identification (Cowan, 1968). In bacterial taxonomy there have been important advances in recent years in all three parts.

In one sense it is impossible to review, with any justice, the detailed progress made in bacterial taxonomy of even just one year. Practically all items of new knowledge gained about the biology of microorganisms have, ultimately, taxonomic significance. After all, bacteria are classified, named and identified on the basis of our total knowledge of them (Rescigno & Maccacaro, 1961; Hill, 1973; but see also Pratt, 1972, 1974).

This chapter will therefore attempt to sketch the trends that have appeared recently in bacterial taxonomy and which are of general interest or major innovation. Technical detail will be avoided: taxonomy is a big enough subject.

## Classification

### Numerical Taxonomy

Numerical taxonomy was first introduced to microbiology in 1959 by Sneath (1959a, b), who has remained the leading exponent of this school. Numerical taxonomy has since become a very familiar, if not routine practice. This wide acceptance has not ossified the subject; on the contrary, there have been several important innovations in recent years which are likely to affect its future development considerably.

## Philosophy

The case for and against the use of numerical taxonomy, will not be argued further here; interested readers may refer to several books now available. The first text book, Sokal and Sneath (1963), has been completely re-written (Sneath & Sokal, 1973). Further books are by Jardine and Sibson (1971), Blackith and Reyment (1971), and Clifford and Stephenson (1975). An outline of the essential elements of numerical taxonomy is given in Skerman (1967) and a thorough introduction in Sneath (1972). Accounts of 'workshop-type' conferences have been published by Cole (1969) and Estabrook (1975). A very extensive bibliography is now available: lists are given in Maisel and Hill, (1969), Crovello and Moss, (1971) and in Sneath and Sokal (1973). It would be dangerous to select particular papers from this extensive bibliography except to mention two particularly interesting papers: Johnson (1968) and Sneath (1971).

## Limitations

Although certain authors were, from the outset, very cautious about the interpretation of the results of their numerical taxonomies, there is now a wider appreciation of the real limitations of these methods. On the one hand, they open the door to much wider studies both in terms of numbers of 'operational taxonomic units' (OTUs, in microbiology usually the strains themselves, Sokal & Sneath, 1963) and in terms of numbers of tests; for the computer can handle quite surprising quantities of data. On the other hand, this very increase in scale of particular studies has led to certain problems that already existed, becoming acute. These concern representation of taxonomic results, questions about reliability of test results and of within- and between-laboratory test reproducibility. These are the subjects of current activity in bacterial classification and should be viewed in toto as a major effort for greater precision in taxonomy in general. This effort could probably not even have been made without the influence of numerical taxonomy.

It is now more widely appreciated that the end product of a numerical taxonomy study is a hypothesis. A numerical taxonomy shows the apparent phenetic relationship (see p. 106), among the OTUs studied in terms of the tests used. As with any hypothesis, it should itself be subject to further test and if needed, change (Silvestri & Hill, 1964).

## Representation of Results

The dendrogram, or taxonomic tree, (a tree-like figure with a similarity scale ordinate indicating the similarity levels at which branches join together; the OTUs are the tips of the branches) remains the most common way of representing results, although it has several defects or limitations. Firstly, when the study contains a few hundred OTUs, the detailed dendrogram becomes almost unmanageable in size, requiring pull-out pages in Journals and more seriously, it may fail in its main objective which is to present ready, visual appreciation of overall relationships. Often authors make finer subdividing steps in their dendrograms than the standard error of the matching (or other similarity) coefficients can justify (Sneath, 1967). The linking together of two, or more, multi-membered branches at a single similarity level is a gross simplification; a more realistic representation is to make less definite or 'fuzzy' connections in which the branches join together over a range of similarity values (Rohlf, 1970).

Secondly, the dendrogram remains essentially a one-dimensional representation, since there is only one scale: the ordinate (the abscissa has no special meaning). Attempts have been made to increase the dimensionality: the quasi-two dimensional dendrograph of McCammon (1968; some meaning is here attached to the abscissa, thus changing the dendrogram to a graph); the three-dimensional trees of Blackith and Reyment (1971, p. 315). Shaded triangular diagrams (after Sneath, 1957b; boxes in the matrix of similarity coefficients between all pairs of OTUs are shaded to different degrees according to the range of coefficients found; groups of high within-group similarities contrast sharply with between-group lower similarities) and the minimum spanning trees of Gower and Ross (1969); (these are special displays showing how all OTUs link together through their nearest neighbours, in a space in which neither the abscissa nor ordinate are scaled), both require an expert eye for their appreciation. 'Taxometric maps' have been developed by Carmichael and Sneath (1969; like geographical maps with similar OTUs located in the same neighbourhoods, distant from dissimilar OTUs). Three-dimensional models were first attempted by Lysenko and Sneath as early as 1959. They can be readily and accurately made when factor analysis methods of one sort or another are used; these are complex statistical methods of analysing multivariate data to discover underlying trends or 'factors', e.g., Principal Component Analysis (Hill et al., 1965) or principal coordinate analysis (Gower, 1966). These methods have not become popular, yet the limitations imposed by dendrograms appear to demand them on occasion.

E

## Uniform Application of Tests

In bacterial studies, a basic problem is uniform applicability of tests to all the OTUs when these differ in a way affecting the execution and reading of tests. The series of papers by Hutchinson and colleagues (the concluding paper was Hutchinson et al., 1969) overcame this problem for aerobic and anaerobic OTUs. Sneath (1968) considered this problem generally and developed some useful special statistics, for example, to allow for slow and fast growing OTUs in one and the same study.

## Test Errors

Arising from identification work and co-operative studies (see below), the effect of test errors on numerical taxonomy is coming under scrutiny. The importance of this work goes beyond the confines of numerical taxonomic methodology itself and acquires significance for all bacterial taxonomic work. (It is worth emphasizing that the motivators of these lines of enquiry are, in the main, numerical taxonomists). There are essentially two aspects to note. Firstly, the determination of actual levels of test reproducibility and, following from that, technical improvements to better those levels and secondly the development of statistics to take these errors into account in numerical taxonomy work. The initial steps have been taken (Sneath & Johnson, 1972; Sneath, 1974b) but this is an area where considerable future activity can be anticipated. It is a problem not to be underestimated: Snell and Lapage (1973) obtained a 5 per cent within-laboratory discrepancy rate for substrate utilization tests; Sneath (1974b) quotes within-laboratory discrepancy rates can, with care, be kept down to 2 per cent. The actual levels will depend to some extent on which particular organisms and tests are being studied. If kept down to less than 5 per cent, test errors have little effect on numerical taxonomies (Sneath & Johnson, 1972). Between-laboratory discrepancies, however, are often much larger, frequently around 15 per cent and sometimes even 30 per cent, for partciular tests (Lapage et al., 1973a). With this knowledge, derived from several sources, evidently numerical taxonomy studies should also test the internal test error rate; without that information, the study will be suspect.

## Co-operative Studies

Following from the above comments, it could be thought futile at present to attempt between-laboratory co-operative numerical taxonomy studies. The emergence of co-operative studies may be considered to be one of the more exciting advances in recent years. The earliest co-operative work was the Streptomyces International Project (Gottlieb, 1961); however, between-laboratory discrepancies were too high for this study to be fruitful. The Pseudomonas Working Party (Sneath & Collins, 1974) investigated tests to improve between-laboratory reproducibility and, indeed, rather than being a classification study of pseudomonads became a study of test reproducibility per se, with some very valuable results. The International Working Group on Mycobacterial Taxonomy (IWGMT) (Wayne et al., 1971; Kubica et al., 1972; Meissner et al., 1974) took a bold step. As with other co-operative studies, a given set of micro-organisms was distributed to all participating laboratories, but each was then allowed to perform their own sets of tests, in their own way. All the results were then pooled, edited in order to delete obvious duplication of results and then analysed by numerical taxonomy. The so-called 'permissive philosophy' of the IWGMT is, essentially, that if A and B organisms are similar to each other, and different from C, then this should be found in any laboratory; what will differ from one laboratory to another will be the precise similarity levels at which their relationships are formed. The pooled data will yield an 'average' result. This approach to cooperative studies allows separate analyses of the results from individual laboratories, for comparison with each other and with the average result. It defers, initially, the question of test reproducibility but, through separate analyses, permits a systematic investigation to identify those particular tests where standardization difficulties are most pressing (Wayne et al., 1974). This same 'permissive philosophy' was the basis for a primitive method to pool published data from different numerical taxonomies (Hill, 1974a).

It could be thought that 'classification by committee' as it were, would be at least uninspired; however, in one of the more recent IWGMT reports (Meissner et al., 1974), a minority view was published as well as the majority view when there was disagreement on interpretation of the results.

Another co-operative study-group, not yet using methods of numerical taxonomy, is concerned with the genus *Moraxella* (Bøvre et al., 1974). More and more of such study groups are likely to be formed in the future.

## Non-Numerical Taxonomy

### Types of Classifications

'Natural' general classification is what all good taxonomists strive for:

but it may mean different things to different taxonomists. My own views on this subject were given in Hill (1972) where the distinction was drawn between phenetic classification which reveals the 'to-day' relationships and for which numerical taxonomy is appropriate; patristic classification (patristic—'of the fathers'—*sensu* Silvestri & Hill, 1964) which reveals the overall genetic relationships and for which techniques of molecular biology are appropriate; and cladistic classification which reveals the pathways of evolutionary lineages relative to time scales. Patristic and cladistic classification are two aspects of phylogenetic classification. (*Note*: Patristic (from the Greek $\pi\alpha\tau\rho$— father) and cladistic (from the Greek $\kappa\lambda\alpha\delta\sigma\zeta$—young shoot or branch) were first used in taxonomy by Cain and Harrison (1960). Patristic was defined as similarity due to common ancestry, not to convergence. Similarities based exclusively on homologous features are patristic similarities, *sensu* Cain and Harrison. Overall similarity (Sneath 1957a) is comprised of similarity due to common ancestry and due to convergence and is called phenetic similarity. Patristic was used by Silvestri and Hill (1964) for similarities in DNA base sequences, assuming that this similarity is due to common ancestry and not derived by convergence. This similarity falls within the accepted criterion for phylogenetic homology, i.e. presenting numerous detailed resemblances which are not functionally necessary). This subject is thoroughly discussed in Sneath and Sokal (1973) and further interesting reading will be found in Heywood and McNeill (1964) and the published volume of the 1974 Society for General Microbiology symposium *Evolution in the Microbial World* (Carlile & Skehel, 1974; especially the first contribution: Sneath, 1974a).

## Cladistic Classification

In the case of bacteria, cladistic classification may ultimately prove to be unresolvable. For example, the wide range of base compositions in the DNA of bacteria suggests first of all that the bacteria, as a whole, are very ancient indeed. Also, the fossil record, such as it is, cannot yield much information. However, the methods of Le Quesne (1969, 1972), put forward in an entomological context and dependent on a logical analysis of which characters appear to be 'uniquely derived', i.e. changed from a '0' score in a numerical taxonomy input matrix to a '1' score, or vice versa, once only during the evolution of the group, has been explored with microbial data (Hill, 1974c, 1975). It may be possible to draw some very general cladistic conclusions, but the likelihood of obtaining detailed cladistic classifications of bacteria is probably very remote; very few of the present day characters of bacteria appear to be 'uniquely derived', at least in Le Quesne's sense.

## Patristic Classification

Considerable impetus has been given, on the other hand, to patristic classification through the study of bacterial genomes by techniques of molecular biology. Work in this area is progressing steadily, although it is probably true to say that innovations in recent years have been mainly at the technical level. Jones and Sneath (1970) reviewed and summarized the main taxonomic groupings of bacteria arising from the study of genomes (see below) and from 'classical' genetic studies which, for bacteria, are mainly phenomena of transduction, transformation and conjugation.

## Genome Sizes

From sedimentation rates under ultracentrifugal conditions together with renaturation rates of denatured DNA, it is possible to determine the molecular weight of the genome, the genome size. This has been found to vary considerably from organism to organism. (De Ley, 1969, 1971a; Gillis et al., 1970; Bak et al., 1970). It would appear reasonable to suppose that organisms with widely different genomic sizes must be patristically distantly related. At the same time, as with DNA base compositions, (see below) the reverse, i.e. possession of similar genomic sizes, need not necessarily imply close patristic affinity. One taxonomic conclusion that has been based on genomic sizes, in conjunction with a phenotypic trait is the classification of the sterol-independent mycoplasmas as a family, the Acholeplasmataceae, separate from the sterol-dependent mycoplasmas, Mycoplasmataceae (Edward & Freundt 1970). The acholeplasmas have a genomic size twice that of the mycoplasmas (Bak et al., 1970).

## Base Composition

Estimation of base compositions of bacterial DNA has become common-place (tabulated data given in Hill, 1966 and Normore, 1973). The so-called 'melting temperature' method of Marmur and Doty (1962) and the buoyant density method of Schildkraut et al., (1962) remain the most popular methods. (Note: one does not 'determine' base composition by these methods; one determines the parameter, melting-temperature or buoyant density, from which base composition is estimated). The melting temperature method has very high within-laboratory reproducibility but, when one considers that base compositions of mutant strains are identical so far as the techniques can detect with parent strains (De Ley, 1964; but see also Weed, 1963), and that DNA base composition is a physico–chemical parameter not subject to variation as a result of growing conditions, the between-laboratory reproducibility is really very bad. A 1°C difference in the melting temperature between laboratories for one and the same DNA sample may not sound very great, but it corresponds to approximately a difference of 2·5 per cent GC in base compositions when calculated with Marmur and Doty's (1962) equation. Up to 5 per cent GC calculated differences can be found in the literature. There are, unfortunately, several possible explanations: the use of a variety of equations relating melting temperatures to base compositions, the use of dilute buffers (Owen et al., 1969), alternative ways of plotting melting curves (Knittel et al., 1968), and even simple errors in calibrating thermometers may be a source of discrepancies (L. R. Hill & H. I. Garvie, unpublished). The conclusion to be drawn is that workers using the melting temperature method to estimate base composition should always include a reference DNA in their series, as is done with buoyant density estimations. The neotype of *E. coli* (NCTC 9001) is a suitable reference organism. Workers should then quote the melting temperature of *E. coli* DNA that was determined under their particular conditions (Mandel et al., 1970).

It is worth repeating, DNA base compositions can strictly only be used to draw negative taxonomic conclusions. When differences are great, then little or no similarity in base sequence, i.e. patristic affinity, can be assumed but when base compositions are similar or identical, this does not necessarily imply close patristic affinity (Hill, 1968). How great is great in this context? Sueoka's (1961) original estimate, from statistical analysis of buoyant density profiles, that a 10 per cent difference in the mean per cent GC value of two DNA samples means virtually no overlap between the two distributions of base compositions, has been revised upwards to 25 per cent differences by De Ley (1969) from the analysis of melting curves.

## Doublet Analysis

Mention must be made of 'doublet analysis': characterization of DNA in terms of the sixteen possible two-base (doublet) sequences along its length. Methods are available to estimate the frequency distribution of each doublet, to compare the actual frequencies with those to be expected if the bases were distributed randomly which is easily calculated from the overall base composition, and even to compare different DNAs of differing base composition in terms of their similar, or dissimilar, deviation from random of their doublet distributions. This work originated as long ago as 1961 (Josse et al., 1961; and Swartz et al., 1962) and has been fruitfully pursued by Subak-Sharpe and colleagues (Subak-Sharpe et al., 1974). From a theoretical standpoint, doublet analysis could acquire considerable taxonomic importance: it is a first step towards determination of DNA base sequences and has direct patristic significance. The results obtained by Subak-Sharpe and his colleagues, especially with reference to viral and bacteriophage nucleic acids, are very interesting. It is doubtful, however, whether this type of work could ever be carried out anywhere other than a specialized biochemistry laboratory.

## Molecular Hybridization

There are also the techniques of in vitro so-called 'molecular hybridization' of single strand DNA from different organisms: if a stable molecular duplex can be formed between single strand DNA from different organisms then, without knowing the actual base sequences, it can be assumed that the base sequences are the same or closely similar. The methods have been reviewed by De Ley (1971b). The early techniques, using the ultracentrifuge (Schildkraut et al., 1961) or agar columns (Bolton & McCarthy, 1962; McCarthy & Bolton, 1963) were soon found inconvenient; or they were not easily made quantitative, or were too laborious, and/or had problems with reproducibility. The introduction of membrane filters (Nygaard & Hall, 1963) was an important advance and permitted 'batch analysis' i.e., simultaneous determination of many reactions. De Ley and Tijtgat (1970) claimed that membrane filters, like agar, suffered, under certain conditions, considerable leaching of supposedly permanently fixed DNA leading to inaccurate results. With these methods, one of the single-strand DNA samples is immobilized either in the agar or on the membrane filter, while the other DNA sample, single-strand and radioactive, is in free solution. The amount of radioactivity that during the experiment also becomes fixed to the supporting material (through duplex formation with the first DNA) is a measure of the molecular homology. De Ley and colleagues (Gillis et al., 1970; De Ley et al., 1970) devised a simple method based on the comparison of renaturation rates of the two DNA samples separately and mixed. They obtained satisfactory results but the method has not come into any wider usage (see however Martin &

Phaff 1973). The most popular method, arising from the earlier work of Walker and his colleagues (Walker & McLaren, 1965) utilizes the differential adsorption of single and double strand DNA on to hydroxy-apatite (Brenner et al., 1972).

In the study of in vitro DNA duplexes it is essential to determine not only whether duplexes are formed but also how much mis-matching of bases is present by determining the melting temperature of the duplex. If the duplex melts at the same temperature as the 'parent' DNAs. then there is no mis-matching; if the melting temperature is lower, then some mis-matching must be present. Brenner et al., (1972) have discussed these methods and concluded that a decrease of 1°C in the melting temperature of the duplex corresponds to approximately 1 per cent mis-matching of bases.

## Nomenclature

Nomenclature is often thought of as the least interesting part of taxonomy or, at best, a tiresome chore. Yet, to use Ainsworth's (1955) description it is the handmaid of classification. It is also the vehicle by which communication in identification is achieved. Its essential role must therefore be taken for granted. There have been some important developments of which even the most disinterested reader must be made aware.

### Bergey's Manual

An eighth edition of *Bergey's Manual* (Buchanan & Gibbons, 1974) has, at last, been published. The time-gap between these editions, 17 years, was too long for it coincided with a period of growing interest in classification from which many nomenclatural changes were bound to come. However, any treatise will always be out of date to some extent. Reference should be made to the International Journal of Systematic Bacteriology (IJSB) as the main Journal to follow to keep up-to-date. It publishes not only taxonomic papers, but also reports, minutes, etc., of the various international taxonomic subcommittees. It is to have a crucial role for nomenclature, even more so in the future (see p. 111).

The new Bergey's Manual differs from previous editions in two important respects. The formal hierarchical structure is taken no further than the family level. Higher taxa (sub-orders, orders) are replaced by informal groupings such as 'Gram-positive cocci'. Secondly, type strains are quoted by Culture Collection numbers for almost all the species listed (*see* also Sneath & Skerman, 1966). DNA base compositions are given when known and much greater use is made of comparative tables and also of locating certain species as 'specie incertis sedis'. Skerman's supplementary *Guide to the Identification of Bacterial Genera* has also been updated and, as with the seventh edition, incorporated in the same volume.

Between the seventh and eighth editions many bacterial names were collected together in a separate volume, *Index Bergeyana* (Buchanan et al., 1966; Addenda by Hatt & Zvirbulis, 1967 and Zvirbulis & Hatt 1969a, b).

## The Bacteriological Code

Bacterial nomenclature is governed by its own Code, comparable to the International Code of Botanical Nomenclature (1969) and the corresponding Zoological Code (1964). There exist other specialized codes; Jeffrey (1973) gives an easy-to-read guide to nomenclatural Codes. At the moment of writing, the International Code of Nomenclature of Bacteria (Editorial Board, 1966) is in force. This 1966 Code replaced the earlier International Code of Nomenclature for Bacteria and Viruses (1958). The nomenclature of viruses is now separate from that of bacteria (Lwoff, 1967; Wildy, 1971 and Fenner et al., 1974). The 1966 Bacteriological Code is itself due to be replaced in 1975 by a new Code. The new code has been published as a Proposed Revision (Lapage et al., 1973b) and was approved, with minor changes, at the first International Congress of Bacteriology, Jerusalem, in 1973 and will become effective 1 January 1976. This will be in book form and will also include the Statutes of the International Committee for Systematic Bacteriology (ICSB) and of the Bacteriology Section of the International Association of Microbiological Societies (IAMS).

Over the years, some dissatisfaction with the 1966 Code had grown and also the ICSB had since approved certain principles (see below). There was a real need, therefore, to revise the 1966 Code, but the new Code does much more than simply tidy these general matters up; it contains some important far-reaching innovations.

Firstly, the starting date for bacterial nomenclature in the 1966 Code was 1753! This has not only been altered but, significantly, put into the future to 1 January, 1980. Between now and 1980, the Taxonomic Subcommittees of the ICSB for the various groups of organisms, and other experts, will draw up 'approved lists of names of bacteria' to be published in the IJSB and which will have 1 January 1980 as their dates of valid publication. This means the experts will carry out a once-and-for-all search of the old literature: nomenclature prior to 1980 can then be conveniently forgotten.

Of course, classification will not stand still after 1980. A second provision of the new Code is that after that date all nomenclatural 'requests for opinions' and new names or combinations of names shall be announced in the IJSB, or at least referred to there if the detail is given elsewhere. Therefore, to keep abreast with future changes, just the one Journal will be needed.

A third provision of the new Code is that the type strain of a named species must be unambiguously designated, and the species itself will have no nomenclatural validity if this is not done. Many nomenclatural disputes arose in the past when arguments were based solely on descriptions, there being no extant type strains available; in Sneath and Skerman's (1966) list of type and reference strains, it is surprising how many even well-known species lacked a type strain. All the names in the eventual approved lists will also have type strains designated. The new Code further recommends that the type strains be deposited in recognized culture collections (see below). There are several different kinds of type strains, e.g., holotypes, neotypes; see Sneath and Skerman (1966) for definitions.

A fourth innovation is that the new Code recommends that 'lists of minimal standards' be drawn up for the various groups, a further task for the Taxonomic Subcommittees. Two such standards have already been published, as proposals; one for *Mycoplasmatales* (Subcommittee on the Taxonomy of Mycoplasmatales, 1972): and one for the genus *Brucella* (Corbel & Brinley Morgan, 1975).

The new Code represents a significant advance in nomenclatural thinking and, in due course, practice. It maintains the original desirable features embodied in the previous Codes, and thus follows on naturally from them. At the same time, the sum effect of the above-mentioned innovations in the new Code will be that, after 1980, only one Journal need be consulted for nomenclatural correctness and only as far back as 1980. The validity of bacterial names will then have precise applications for they will be associated with descriptions of at least a minimal standard and with extant type strains.

## Culture Collections

Though not strictly coming under the heading of 'nomenclature', it will be convenient to mention here the organizational innovations of recent years in connection with culture collections. Culture Collections have long played an important role in the development of microbiology.

Various international agencies have come to recognize this fact in post-war years. In 1970 the World Federation for Culture Collections (WFCC) was set up (Martin, 1972; Lapage, 1972) replacing the earlier established Section on Culture Collections, a section of IAMS. The WFCC is a 'Commission' within each of the three Divisions of Microbiology, Botany, and Zoology of IUBS (International Union of Biological Sciences), and a 'Federation' within IAMS. The aims and purposes of the WFCC are contained in its Statutes (World Federation for Culture Collections: Statutes, 1972), and a World Directory of Culture Collections has been compiled (Martin & Skerman, 1972). A general discussion on culture collections and the WFCC will be found in Lapage (1971). As more than 300 culture collections throughout the world are listed in the Directory, there can be little excuse for not depositing type strains, as required by the new Code.

The WFCC is also involved in collecting data on the reactions of deposited strains. This is a mammoth task and data centres have been set up, or are in the process of being organized, in Australia, (Skerman, 1973), the United States, and possibly in Japan. Rogosa et al., (1971) have proposed a comprehensive scheme for uniform coding of data. Uniform coding is a severe problem, of course, despite the prowess of computers (*see* Krichevsky & Norton, 1974, for a general discussion). Comparability of test results depends on the use of the same testing method and sometimes on the precise way tests are carried out. Standardization of test procedures is a difficult problem but Skerman (1969) has taken a first step by compiling into one book a vast number of test procedures known to be in use.

## Identification

The identification of higher plants and animals is usually made by visual examination of the specimen and simultaneous use of an identification scheme or key, traditionally contained in a book or, now coming into usage, stored in a computer. Several 'on-line' computer programmes for this kind of question-and-answer identification have been described and the most advanced systems allow several items of information about the specimen to be supplied simultaneously or in no particular sequence (Goodall, 1968; Morse, 1971; Pankhurst, 1974, 1975). In bacterial identification, books such as *Bergey's Manual* or Skerman (1967) are of general utility, while for medical bacteriology, those of Cowan and Steel (1965), largely re-written by Cowan (1974), and *Topley and Wilson* (new edition, Wilson & Miles, 1975) are more useful. However, simultaneous examination and identification, which is practical with plants and animals, is not possible with bacteria since it takes time to set up and read the results of tests to obtain the necessary information. Bacterial identification is a relatively slow process and to improve on the traditional methods advances are being made on two fronts: data acquisition and data processing.

## Data Acquisition

*Specialized Apparatus.* Microbiologists have always sought to improve the conventional methods of examining cultures, e.g., Clarke and Cowan's (1952) micro-tests, or divided petri dishes (Stevens, 1969). Some recent innovations are known to work well and to be time-saving for a particular type of test, e.g., the electrical impedance method for rapid antibiotic assays (Ur & Brown, 1975). Trotman (1972) gave a comprehensive review of the many specialized technical innovations of recent years leading to automation of laboratory test procedures (*see* also Baillie & Gilbert, 1970; and Hedén & Ilteni, 1974). Most of these devices are, however, designed to automate a particular test or narrow range of tests, but an important common feature to many of them is that they often achieve not only automation per se, but also improve the test to a higher level of quantitative measurement.

*Test kits.* The introduction of various commercial 'test-kits' is another important development. Many of these are now available for the commoner groups of organisms (*see* Nord et al., 1974 for a comparison of the API, AuxoTab, Enterotube, Pathotec and R/B systems), The various kits achieve some or all of several objectives: miniaturizing conventional tests, thus allowing the main body of data accumulated over the years regarding test-responses of taxa to be utilized (but *see* p.115); factory production, dispensing and sometimes dehydration of the various media, leading hopefully, to greater consistency in media compositions; and arrangement into single inoculating units. Considerable economy in materials and technicians' time is gained.

Test-kits are already in use successfully for routine identifications in clinical laboratories. Most were initially developed to identify members of the *Enterobacteriaceae*; they began modestly enough (e.g., API-20, for only twenty tests) but some are rapidly expanding in terms of numbers of tests and range of organisms for which they can be used, e.g., Duborgel and Chouteau (1974) recently reported the satisfactory performance of an API-50 test-kit to identify anaerobic organisms.

This type of practical improvement in bacterial identification is, I feel, a very real advance. The efficiency of test kits depends heavily on the initial choice of the correct set of tests to include in the system and, it is fair to point out, conventional methods retain the advantage of a flexible choice of which tests to use. Consequently, the manufacturers must spend considerable effort to choose 'best sets' which requires knowledge gained from processing much data. The current API systems are supplemented by a companion ready-reckoner type of identification scheme.

It does not really matter very much if results obtained in these systems do not correspond exactly with those that would be obtained by conventional means. Lack of correspondence does not affect the value of the system; it merely means that the criteria for identification may require modification.

A possible limitation of test kits is their capacity to identify correctly unusual organisms, of which there are two main kinds: first, the rare, unexpected or little known organism, e.g., *Eikenella corrodens* which is not difficult to identify, and, second, a strain that gives one or more reactions atypical of its species. The success of the API and similar systems in the identification of unusual organisms will depend on the number and choice of tests included in the set. While an expanded test kit should be more capable than a restricted one, yet at the same time, a major motivation for test kits in the first place is to achieve economies. Since the commoner organisms, which comprise the bulk of routine identifications, can be adequately identified with the smaller sets, there appears to be a conflict of objectives here. The danger to be recognized is that too much reliance on especially the smaller systems would require a very alert bacteriologist indeed to spot the unusual cases which require a more thorough examination.

*Complete Automation.* A fundamentally different approach to improving data acquisition is to seek new types of information with automation in mind from the outset. The departure is not to modify conventional tests but ultimately to devise 'black boxes' in which organisms enter on one side and answers emerge on the other. The most important development in this direction is the application of gas chromatographic techniques to analyse total chemical composition, fractions of cells, or metabolic products. The literature on this subject has grown considerably but there seems to be no suitable review article to quote and to do so here would require a chapter to itself. Simply as a recent example of a particular gas liquid chromatography (GLC) application, Jantzen et al., (1974) analysed the fatty acids of whole-cell methanolysates and applied numerical taxonomy to the results. Mention must also be made of pyrolysis-GLC methods pioneered by Reiner and his colleagues

(Reiner, 1965; Reiner & Ewing, 1968; Reiner et al., 1972, 1973). The output in this case is a profile of the total chemical composition of the organisms.

Profiles, either total or partial, from GLC applications are, like finger-prints (and indeed sometimes referred to in the literature as such) different for different species. Strain differences in profiles must be smaller than species differences, computer methods are needed to com-pare profiles and, of course, an adequate library of reference profiles must be built up. There are also reproducibility problems, particularly the effect on the profile of the growth medium. Nevertheless, the output from these methods is suitable for computer processing, thus opening the door to completely automated identification systems.

## Data Processing

Despite the wide application of numerical taxonomy to classification, with a large literature now accumulated, the parallel development of numerical identification has been slow. A reason for this may be that for classification work the taxonomist collects data in a relatively leisurely way, carefully compiles the input and then has the data processed as a 'one-off job'; he needs only occasional access to the computer. With identification, the need is for frequent computer access. Time-sharing computer systems are now well-developed with the result that direct access is readily available; only a remote computer terminal and tele-phone link is required. With this development, recent years have seen increased interest in numerical identification.

*Probabilistic Nature of Identification.* Numerical identification can be achieved in several ways (Hill, 1974b, for a general discussion) but it is worth re-emphasizing the probabilistic nature of identification. In an absolute sense, any identification is probabilistic: strains are allocated to previously defined taxa, but these themselves are part of a classification which is an hypothesis. The classificatory context within which identifi-cation proceeds is not fixed or final, but is subject to change. Identification is usually made on fewer characters than used to devise classifications, so is carried out in partial ignorance. There is always the possibility that the unidentified strain belongs to a new taxon not included in the existing classification. These general constraints are applicable to all identification work but methods of numerical identification need to take particular attention of them. Numerical methods allow measures of probability to be calculated but there is a risk that in so doing the more general and basic probability may be overlooked.

*Data Reduction.* The first approach to numerical identification is to choose from the number of tests required for classification a smaller number necessary for identification. Once groups have been formed and defined in a classification, numerical methods can be used to extract the minimal, or near minimal, amount of information necessary to separate the groups. These methods use various measures of the 'information content' of tests, originally defined by Rescigno and Maccacaro (1961). Several mathematical measures of the information content of tests have since been proposed: Möller (1962), Gyllenberg (1963), Rypka et al., (1967), Niemela et al., (1968), Pankhurst (1970), Gower and Barnett (1971); summarized and compared in Hill (1974b). Some of these measures allow for missing data or consider tests in pairs, or threes, etc., instead of singly. By these measures, tests can be ranked in order of decreasing importance for identification and either an identification matrix drawn up or they can be arranged in a sequential key for identification. Sequential keys can be made statistically probabilistic by allocating probabilities at each bifurcation (Möller, 1962; Hill & Silvestri, 1962), though this approach has not been much used.

*Probability Matrices.* A more comprehensive approach than keys or reduced matrices is the method of transforming taxa versus test result matrices in toto into probabilistic identification matrices, by replacing original '+' or '−' signs by probability figures. Unknowns are then compared with all the taxa in the matrix with the comparison based on those tests for which the reactions of the unknown have been recorded. The corresponding probability figures are multiplied, taxon by taxon, and the most likely taxon for the identification found. The original method was devised by Dybowski and Franklin (1968); it has been fully developed by Lapage and colleagues (Lapage et al., 1973a; Lapage, 1974). Their current system is very comprehensive from the practical point of view (Bascomb et al., 1973), and has a sound theoretical basis (Willcox et al., 1973). Limits to the number of tests and of taxa contained in the identification matrix are set only by practical considerations such as the capacity of the computer. The information available about the unknown is used and the organism identified or not identified. If identified, then this is with a certain probability of error which is fixed, at present, at one in a thousand, for Gram-negative organisms; with other groups less stringent acceptance levels may have to be set. If not identified remaining tests in the matrix will be examined to determine which tests ought to be carried out to further the identification, maximizing the chances of success. If these further tests are carried out on the unknown, then all the data can be re-analysed. The print-outs often have some other useful information, e.g. unusual results given by the unknown for the taxon with which it has been identified.

The above-quoted papers of Lapage and colleagues describe the various technical problems and the difficult problem of how to assess performance of the method. They report, on the basis of processing many hundreds of strains, very high identification rates when the reference data is adequate: 85–90 per cent success for fermentative Gram-negative organisms. Identification rates fall dramatically when the reference data is poor: 80 or 50 per cent dependent on whether culture collection deposited strains or field strains were used to test the system, for non-fermentative Gram-negative organisms. The lower identification rate, however, was obtained with field strains that had been submitted for identification as difficult strains selected from the many thousands of strains examined in diagnostic laboratories. An improved identification matrix for non-fermentative organisms has since been compiled (S. P. Lapage, *personal communication*). Even if initially the method should give 'poor' results (which, in fact, may be no worse than conventional identification) with particular taxa or groups of taxa, an advantage of the method is that its use will provide the information necessary to improve performance. Once in use, the system can be readily improved.

The probabilistic identification matrix method of Lapage and his colleagues does not specifically require a numerical taxonomy to be carried out first. Provided that there is sufficient data available in the literature on the test responses of taxa to permit probability figures to be allocated, then the method can proceed.

*Alternative Methods.* The probabilistic identification matrix method is more flexible than several alternative approaches that have been suggested. Alternative methods which also use computers to aid identification include the comparison of unknowns with hypothetical or real median organisms, or with centrotypes (OTUs with the highest average similarity with all the other OTUs in the same group or cluster) or centres of gravity of taxonomic clusters. Median organisms can easily be computed as an end-product of a numerical taxonomy (Liston et al., 1963; *see* also Tsukamura & Mizuno, 1968) and centrotypes readily chosen (Silvestri et al., 1962). Gyllenberg (1965) suggested that, in factor analysis types of numerical taxonomy, centres of gravity of clusters could be calculated and unknowns compared with these. However, these methods require that a numerical taxonomy be carried out first.

Lapage and Willcox (1974) have suggested a simple manual method of analysing data whereby the typical reactions of taxa are quickly found and aberrant strains readily detected.

Finally, the importance of test errors on all these methods of numerical identification has to be re-emphasized. The papers of Sneath and Johnson (1972), Sneath (1974b), and those of Lapage and his colleagues have already been mentioned. Much of the concern, and consequent work, with this aspect has arisen as a result of the endeavour to achieve successful numerical identification as a practical proposition in microbiology.

## Conclusion

In all three of its parts, bacterial taxonomy is today very much alive. In the long term, all branches of microbiology may pass through evident phases of activity and inactivity, and taxonomy is no exception. Although taxonomy is one branch of continual improvement and revision, the current level of activity amounts to much more than 'routine' improvements: they amount to a revolution that, without knowing what the future will hold, will mark the current phase as very significant.

In classification, the major recent innovations have been in numerical taxonomy which is now undertaking the more difficult problems (test reproducibility and collaborative work), solutions to which will affect all taxonomic work. Innovation in the molecular biology area is mainly technical, but these methods too are coming into wider usage; the techniques are demanding and consequently their application to taxonomy results in higher levels of laboratory expertise.

The effect on nomenclature of these developments in classification may or may not be great; we shall have to wait and see. But independently, nomenclature too is entering a new phase arising from the innovations contained in the revised Code. The implications of the new Code mean considerable work ahead for taxonomists, but for the non-taxonomist user they should result in a great simplification of the procedures necessary to ensure correct nomenclature.

The developments towards automated identification, at both data acquisition and data processing levels, are gathering momentum; methods of computer identification have had their practical utility demonstrated and only a widening of such systems can be envisaged. The linkage of data acquisition and data processing has begun.

These advances are changing, or have already changed, the character of bacterial taxonomy. Long thought of as merely the descriptive part of microbiology, or even considered by some to be more an art than a science, bacterial taxonomy is now an experimental science.

What can be said of the effect of these changes on taxonomists themselves? There may linger a suspicion that, by giving greater roles to the computer in all parts of taxonomy, the taxonomist may become little more than a press-button mechanic. The contrary is true of course, for taxonomy is a more demanding subject than ever before and the taxonomists' role that much more crucial. This goes beyond the adage about computers of 'garbage in, garbage out', with the corollary that the taxonomist must ensure that there is no garbage in the first place; what has happened is that the taxonomic horizon itself has been considerably widened and the tasks that have become possible and practical are that much larger.

## Acknowledgement

I am grateful to Dr S. P. Lapage for his advice and critical reading of the manuscript.

## References

Ainsworth, G. C. (1955) Nomenclature, the handmaid of classification. *J. gen. Microbiol.* **12**, 322.

Baillie, A. & Gilbert, R. J. (1970) *Automation, mechanization and data handling in microbiology.* The Society for Applied Bacteriology, Technical Series No. 4. 233 pp. London: Academic Press.

Bak, A. L., Black, F. T., Christiansen, C. & Freundt, E. A. (1970) Genome size of mycoplasmal DNA. *Nature, Lond.,* **224**, 1209.

Bascomb, S., Lapage, S. P., Curtis, M. A. & Willcox, W. R. (1973) Identification of bacteria by computer: identification of reference strains. *J. gen. Microbiol.,* **77**, 291.

Blackith, R. E. & Reyment, R. A. (1971) *Multivariate Morphometrics.* 412 pp. London: Academic Press.

Bolton, E. T. & McCarthy, B. J. (1962) A general method for the isolation of RNA complementary to DNA. *Proc. Natn. Acad. Sci. N. Y.,* **48**, 1390.

Bøvre, K., Fuglesang, J. E., Henriksen, S. D., Lapage, S. P., Lautrop, H. & Snell, J. J. S. (1974) Studies on a collection of Gram-negative bacterial strains showing resemblance to Moraxellae: Examination by conventional bacteriological methods. *Int. J. Systemat. Bacteriol.,* **24**, 438.

Breed, R. S., Murray, E. G. D. & Smith, N. R. (1957). *Bergey's Manual of Determinative Bacteriology,* 7th ed. 1094 pp. London: Baillière, Tindall and Cox.

Brenner, D. J., Fanning, G. R., Skerman, F. J. & Falkow, S. (1972) Polynucleotide sequence divergence among strains of *Escherichia coli* and closely related organisms. *J. Bacteriol.,* **109**, 953.

Buchanan, R. E. & Gibbons, N. E. (Eds) (1974) *Bergey's Manual of Determinative Bacteriology,* 8th ed. 1246 pp. Baltimore: Williams & Wilkins Co.

Buchanan, R. E., Holt, J. G. & Lessel, E. F. Jr. (Eds) (1966) *Index Bergeyana.* 1472 pp. London: E & S. Livingstone Ltd.

Cain, A. J. & Harrison, G. A. (1960) Phyletic weighting. *Proc. zool. Soc., Lond.,* **135**, 1.

Carlile, M. J. & Skehel, J. J. (Eds) (1974) *Evolution in the Microbial World.* 24th Symposium of the Society for General Microbiology, 430 pp. Cambridge: Cambridge University Press.

Carmichael, J. W. & Sneath, P. H. A. (1969) Taxometric maps. *Systemat. Zool.,* **18**, 402.

Clarke, P. H. & Cowan, S. T. (1952) Biochemical methods for bacteriology. *J. gen. Microbiol.,* **6**, 187.

Clifford, H. T. & Stephenson, W. (1975) *An Introduction to numerical classification,* London: Academic Press.

Cole, A. J. (Ed.) (1969) *Numerical Taxonomy,* 324 pp. London: Academic Press.

Corbel, M. J. & Brinley Morgan, W. J. (1975) Proposal for minimal standards for descriptions of new species and biotypes of the genus *Brucella. Int. J. Systemat. Bacteriol.,* **25**, 83.

Cowan, S. T. (1968) *A Dictionary of Microbial Taxonomic Usage,* 118 pp. Edinburgh: Oliver & Boyd.

Cowan, S. T. (1974) *Cowan and Steel's Manual for the Identification of Medical Bacteria.,* 238 pp. Cambridge: Cambridge University Press.

Cowan, S. T. & Steel, K. J. (1965) *Manual for the Identification of Medical Bacteria,* 217 pp. Cambridge: Cambridge University Press.

Crovello, I. J. & Moss, W. W. (1971). A bibliography on classification in diverse disciplines. *Classification Soc. Bull.,* **2**, 29.

De Ley, J. (1964). Effect of mutation on DNA-composition of some bacteria. *Antonie van Leeuwenhoek,* **30**, 281.

De Ley, J. (1969) Compositional nucleotide distribution and theoretical prediction of homology in bacterial DNA. *J. theoret. Biol.* **22**, 89.

De Ley, J. (1971a) The determination of the molecular weight of DNA per bacterial nucleoid. *In: Methods in Microbiology,* J. R. Norris & D. W. Ribbons (Eds), Volume 5A. pp 303–309. London: Academic Press.

De Ley, J. (1971b) Hybridization of DNA. *In: Methods in Microbiology.* J. R. Norris and D. W. Ribbons (Eds), Volume 5A. pp 311–329. London: Academic Press.

De Ley, J., Cattoir, H. & Reynaerts, A. (1970) The quantitative measurement of DNA hybridization from renaturation rates. *Euro. J. Biochem.,* **12**, 133.

De Ley, J. & Tijtgat, R. (1970) Evaluation of membrane filter methods for DNA-DNA hybridization. *Antonie van Leeuwenhoek,* **36**, 461.

Duborgel, S. & Chouteau, J. (1974) The advantage of a new micromethod for the identification of anaerobes: a comparison with the conventional system. *XIth Conference on the Taxonomy of Bacteria, Brno, Czechoslovakia:* Abstracts, No. 6.

Dybowski, W. & Franklin, D. (1968) Conditional probability and the identification of bacteria: a pilot study. *J. gen. Microbiol.,* **54**, 215.

Editorial Board (1966) *International Code of Nomenclature of Bacteria. Int. J. Systemat. Bacteriol.* **16**, 459.

Edward, D. G. ff. & Freundt, E. A. (1970) Amended nomenclature for strains related to *Mycoplasma laidlawii. J. gen. Microbiol.* **62**, 1.

Estabrook, G. F. (Ed.) (1975) *Proceedings of the Eighth Annual International Conference on Numerical Taxonomy.* San Francisco: W. H. Freeman (In Press).

Fenner, F., Pereira, H. G., Porterfield, J. S. Joklick, W. K. & Downie, A. W. (1974) Family and generic names for viruses approved by the International Committee on Taxonomy of Viruses, June 1974. *Intervirology*, **3**, 193.

Gillis, M., De Ley, J. & De Cleene, M. (1970) The determination of molecular weight of bacterial genome DNA from renaturation rates. *Euro. J. Biochem.*, **12**, 143.

Goodall, D. W. (1968) Identification by computer. *Bioscience*, **18**, 485.

Gottlieb, D. (1961) An evaluation of criteria and procedures used in the description and characterization of the streptomycetes. A cooperative study. *Appl. Microbiol.* **9**, 55.

Gower, J. C. (1966) Some distance properties of latent root and vector methods used in multivariate analysis. *Biometrika*, **53**, 325.

Gower, J. C. & Barnett, J. A. (1971) Selecting tests in diagnostic keys with unknown responses. *Nature, Lond.*, **232**, 491.

Gower, J. C. & Ross, G. J. S. (1969) Minimum spanning trees and single linkage cluster analysis, *Appl. Statist.*, **18**, 54.

Gyllenberg, H. G. (1963). A general method for deriving determination schemes for random collections of microbial isolates. *Annales Academiae Scientarium Fennicae A* IV **69**, 1.

Gyllenberg, H. G. (1965) A model for computer identification of microorganisms. *J. gen. Microbiol.*, **39**, 401.

Hatt, H. D. & Zvirbulis, E. (1967) Status of names of bacterial taxa not evaluated in Index Bergeyana (1966). I. Names published *circa* 1950–1967 exclusive of the genus *Salmonella*. *Int. J. Systemat. Bacteriol.*, **17**, 171.

Heden, C.-G. & Ilteni, T. (Eds) (1974) *Automation in Microbiology and Immunology. New Approaches to the Identification of Microorganisms*, (2 vols), New York: Wiley.

Heywood, V. H. & McNeill, J. (Eds.) (1964) *Phenetic and Phylogenetic Classification*. Publication No. 6., 164 pp., London: The Systematics Association.

Hill, L. R. (1966) An index to deoxyribonucleic acid base compositions of bacterial species. *J. gen. Microbiol.*, **44**, 419.

Hill, L. R. (1968) The determination of deoxyribonucleic acid base compositions and its application to bacterial taxonomy. *In: Identification Methods for Microbiologists*. B. M. Gibbs and D. A. Shapton (Eds.). *Part B.* p. 177., London: Academic Press.

Hill, L. R. (1972) Prospectives for *Mycoplasma* classification using multivariate analysis methods. *Med. Microbiol. Immunol.*, **157**, 101.

Hill, L. R. (1973) A fault in Pratt's critique of numerical taxonomy? *J. Theoret. Biol.*, **40**, 397.

Hill, L. R. (1974a) Interlocking numerical taxonomies. *XI Conference on the Taxonomy of Bacteria, Brno, Czechosolvakia:* Abstracts, No. 1. (Full text: *Internat. J. Systemat. Bacteriol.*, **25**, 245.

Hill. L. R. (1974b) Theoretical aspects of numerical identification. *Int. J. Systemat. Bacteriol.*, **24**, 494.

Hill, L. R. (1974c) Deduction of phylogenetic (cladistic) groups from phenetic data: application to *Mycoplasma* and *Acholeplasma* spp. *Proc. Soc. gen. Microbiol.*, **1**, 64.

Hill, L. R. (1975) Problems arising from some tests of Le Quesne's concept of uniquely derived characters. *In:* Estabrook, 1975 (*q.v.*) (In Press).

Hill, L. R. & Silvestri, L. G. (1962) Quantitative methods in the systematics of actinomycetales. III. The taxonomic significance of physiological-biochemical characters and the construction of a diagnostic key. *G. Microbiol.,* **10,** 1.

Hill, L. R., Silvestri, L. G., Ihm, P., Farchi, G. & Lanciani, P. (1965) Automatic classification of staphylococci by principal-component analysis and a gradient method. *J. Bacteriol.,* **89,** 1393.

Hutchinson, M., Johnstone, K. I. & White, D. (1969) Taxonomy of the genus *Thiobacillus*: the outcome of numerical taxonomy applied to the group as a whole. *J. gen. Microbiol.,* **57,** 397.

*International Code of Botanical Nomenclature,* (1969) adopted by the Eleventh International Botanical Congress, Seattle, August, 1969. (Utrecht: International Bureau of Plant Taxonomy and Nomenclature. 1972. 426 pp.)

*International Code of Nomenclature for Bacteria and Viruses.* (1958) Ames, Iowa: Iowa State College Press. 186 pp.

*International Code of Zoological Nomenclature,* (1964) adopted by the XV International Congress of Zoology (International Commission on Zoological Nomenclature, London, 1964, 176 pp.)

Jantzen, E., Bergan, T. & Bøvre, K. (1974) Gas chromatography of bacterial whole cell methanolysates. *Acta path. microbiol., Scand.,* Section B. **82,** 785.

Jardine, N. & Sibson, R. (1971) *Mathematical Taxonomy.* 286 pp. London: Wiley.

Jeffrey, C. (1973) *Biological Nomenclature.* 69 pp. Special Topics in Biology Series, The Systematics Association. London: Edward Arnold.

Johnson, L. A. S. (1968) Rainbow's end: the quest for an optimal taxonomy. *Proc. Linn. Soc., New South Wales,* **93,** 8. (Reprinted with additional comments in *Systemat. Zool.,* **19,** 203).

Jones, D. & Sneath, P. H. A. (1970) Genetic transfer and bacterial taxonomy. *Bacteriol. Revs.,* **34,** 40.

Josse, J., Kaiser, A. D. & Kornberg, A. (1961) Enzymatic synthesis of deoxyribonucleic acid. VIII. Frequencies of nearest-neighbour base sequences in deoxyribonucleic acid. *J. biol. Chem.,* **236.** 864.

Knittel, M. D., Black, C. H., Sandine, W. E. and Fraser, D. K. (1968) Use of normal probability paper in determining thermal melting values of deoxyribonucleic acid. *Can. J. Microbiol.,* **14,** 239.

Krichevsky, M. I. & Norton, L. M. (1974) Storage and manipulation of data by computers for determinative bacteriology. *Int. J. Systemat. Bacteriol.,* **24.** 524.

Kubica, G. P., Baes, I., Gordon, R. E., Jenkins, P. A., Kwapinski, J. B. G., McDurmont, C., Pattyn, S. R., Saito, H., Silcox, V., Stanford, J. L., Takeya, K. & Tsukamura, M. (1972) A co-operative numerical analysis of rapidly growing mycobacteria. *J. gen. Microbiol.,* **73,** 55.

Lapage, S. P. (1971) Culture collections of bacteria. *Biol. J. Linn. Soc.,* **3,** 197.

Lapage, S. P. (1972) World Federation for Culture Collections: Xth International Congress for Microbiology: Minutes of the Extraordinary Meeting of the Provisional Board. *Int. J. Systemat. Bacteriol.,* **22,** 404.

Lapage, S. P. (1974) Practical aspects of probabilistic identification of bacteria. *Int. J. Systemat. Bacteriol.,* **24,** 500.

Lapage, S. P., Bascomb, S., Willcox, W. R. & Curtis, M. A. (1973a) Identification of bacteria by computer: general aspects and perspectives. *J. gen. Microbiol.,* **77,** 273.

Lapage, S. P., Clark, W. A., Lessel, E. F., Seeliger, H. P. R. & Sneath, P. H. A. Drafting Committee. (1973b) Proposed Revision of the International Code of Nomenclature of Bacteria. *Int. J. Systemat. Bacteriol.*, **23**, 83.

Lapage, S. P. & Willcox, W. R. (1974) A simple method for analysing binary data. *J. gen. Microbiol.*, **85**, 376.

Le Quesne, W. J. (1969) A method of selection of characters in numerical taxonomy. *Systemat. Zool.*, **18**, 201.

Le Quesne, W. J. (1972) Further studies based on the uniquely derived character concept. *Systemat. Zool.*, **21**, 281.

Liston, J., Weibe, W., & Colwell, R. R. (1963) Quantitative approach to the study of bacterial species. *J. Bacteriol.*, **85**, 1061.

Lwoff, A. (1967). Principles of classification and nomenclature of viruses. *Nature, Lond.*, **215**, 13.

Lysenko, O. & Sneath, P. H. A. (1959) The use of models in bacterial classification. *J. gen. Microbiol.*, **20**. 284.

Maisel, H. & Hill, L. R. (1969) *A KWIC index of publications in numerical taxonomy in the period 1948–1968*. 59 pp. Washington D.C.: Georgetown University Computation Center.

Mandel. M., Igambi, L., Bergendahl, J., Dodson, M. L. Jr. & Scheltgen, E. (1970) Correlation of melting temperature and cesium chloride buoyant density of bacterial deoxyribonucleic acid. *J. Bacteriol.*, **101**, 333.

Marmur, J. & Doty, P. (1962) Determination of the base composition of deoxyribonucleic acid from its thermal denaturation temperature. *J. Molec. Biol.*, **5**, 109.

Martin, S. M. (1972) International Association of Microbiological Societies Section on Culture Collections: Minutes of the Meeting. *Int. J. Systemat. Bacteriol.*, **22**, 406.

Martin, S. M. & Skerman, V. B. D. (Eds.) (1972) *World Directory of Collections of Cultures of Micro-organisms*. 560 pp. New York: Wiley Interscience.

Martini, A. & Phaff, H. J. (1973) The Optical determination of DNA–DNA homologies in Yeasts. *Annali. Microbiol.*, **23**, 59.

McCammon, R. B. (1968) The dendrograph: a new tool for correlation. *Geol. Soc. Amer. Bull.*, **79**, 1663.

McCarthy, B. J. & Bolton, E. T. (1963) An approach to the measurement of genetic relatedness among organisms. *Proc. Natn. Acad. Sci., New York*, **50**, 156.

Meissner, G., Schröder, K. H., Amadio, G. E., Anz, W., Chaparas, S., Engel, H. W. B., Jenkins, P. A., Käppler, W., Kleeberg, H. H., Kubala, E., Kubin, M., Lauterbach, D., Lind, A., Magnusson, M., Mikova, ZD., Pattyn, S. R., Schaefer, W. B., Stanford, J. L., Tsukamura, M., Wayne, L. G., Willers, I. & Wolinsky, E. (1974) A co-operative numerical analysis of nonscoto- and nonphoto-chromogenic slowly growing mycobacteria. *J. gen. Microbiol.*, **83**, 207.

Möller, F. (1962) Quantitative methods in the systematics of actinomycetales. IV. The theory and application of a probabilistic identification key. *G. Microbiol.*, **10**, 29.

Morse, L. E. (1971) Specimen identification and key construction with time-sharing computers. *Taxon*, **20**, 269.

Niemelä, S. I., Hopkins, J. W. & Quadling, C. (1968) Selecting an economical binary test battery for a set of microbial cultures. *Can. J. Microbiol.*, **14**, 271.

Nord, C-E., Lindberg, A. A. & Dahlbäck, A. (1974) Evaluation of five test-kits—API, AuxoTab, Enterotube, PathoTec and R/B—for identification of Enterobacteriaceae. *Med. Microbiol. Immunol.*, **159**, 211.

Normore, W. M. (1973) Guanine-plus-cytosine (GC) composition of the DNA of bacteria, fungi, algae and protozoa. *In Handbook of Microbiology*, A. I. Laskin & H. A. Lechevalier (Eds.), Vol. II, p. 585. Microbial Composition. Cleveland: CRC Press.

Nygaard, A. P. & Hall, B. D. (1963) A method for the detection of RNA-DNA complexes. *Biochem. & Biophys. Res. Commun.*, **12**, 98.

Owen, R. J., Hill, L. R. & Lapage, S. P. (1969) Determination of DNA base compositions from melting profiles in dilute buffers. *Biopolymers*, **7**, 503.

Pankhurst, R. J. (1970) A computer program for generating diagnostic keys. *Computer J.*, **13**, 145.

Pankhurst, R. J. (1974) Automated identification in systematics. *Taxon*, **23**, 45.

Pankhurst, R. J. (Ed) (1975) *Biological Identification with Computers*. Special Topics in Biology Series, No. 7. p. 333. The Systematics Association. London: Academic Press.

Pratt, V. (1972) Numerical taxonomy—A critique. *J. theoret. Biol.*, **36**, 581.

Pratt, V. (1974) Numerical taxonomy: on the incoherence of its rationale. *J. Theoret. Biol.*, **48**, 497.

Reiner, E. (1965) Identification of bacterial strains by pyrolysis-gas-liquid chromatography. *Nature, Lond.*, **206**, 1272.

Reiner, E. & Ewing, W. H. (1968) Chemotaxonomic studies of some Gram-negative bacteria by means of pyrolysis-gas-liquid chromatography. *Nature, Lond.*, **217**, 191.

Reiner, E., Hicks, J. J. Ball, M. M. and Martin, W. J. (1972) Rapid characterization of salmonella organisms by means of pyrolysis-gas-liquid chromatography. *Anal. Chem.*, **44**, 1058.

Reiner, E., Hicks, J. J. & Sulzer, C. R. (1973) Leptospiral taxonomy by pyrolysis-gas-liquid chromatography. *Can. J. Microbiol.*, **19**, 1203.

Rescigno, A. and Maccacaro, G. A. (1961) The information content of biological classifications. *In: Information Theory.* C. Cherry (Ed.) p. 437. London: Butterworth.

Rogosa, M., Krichevsky, M. I. & Colwell, R. R. (1971) Methods for coding data on microbial strains for computers (Edition AB). *Int. J. Systemat. Bacteriol.*, **21**, Special Report A1.

Rohlf, F. J. (1970) Adaptive hierarchical clustering schemes. *Systemat. Zool.*, **19**, 58.

Rypka, E. W., Clapper, W. E., Bowen, I. G. & Babb, R. (1967) A model for the identification of bacteria. *J. gen. Microbiol.*, **46**, 407.

Schildkraut, C. L., Marmur, J. & Doty, P. (1961) The formation of hybrid DNA molecules and their use in studies of DNA homologies. *J. Molec. Biol.*, **3**, 595.

Schildkraut, C. L., Marmur, J. & Doty, P. (1962) Determination of the base composition of deoxyribonucleic acid from its buoyant density in CsCl. *J. Molec. Biol.*, **4**, 430.

Silvestri, L. G. & Hill, L. R. (1964) Some problems of the taxometric approach. *In: Phenetic and Phylogenetic Classification.* V. H. Heywood & J. McNeill (Eds), p. 87. Publication No. 6. London: The Systematics Association.

Silvestri, L. G., Turri, M., Hill, L. R. & Gilardi, E. (1962) A quantitative approach to the systematics of actinomycetes based on overall similarity. *In*: *Microbial Classification*. G. C. Ainsworth & P. H. A. Sneath (Eds), pp. 333–360. 12th Symposium of the Society for General Microbiology. Cambridge: Cambridge University Press.

Skerman, V. B. D. (1967) *A Guide to the Identification of the Genera of Bacteria*. 303 pp. Baltimore: Williams & Wilkins Company.

Skerman, V. B. D. (1969) *Abstracts of Microbiological Methods*. 883 pp. New York: Wiley Interscience.

Skerman, V. B. D. (1973) Statement on the WFCC Center for storage, retrieval, and classification of data on microorganisms. *Int. J. Systemat. Bacteriol.*, **23**, 477.

Sneath, P. H. A. (1957a) Some thoughts on bacterial classification. *J. gen. Microbiol.*, **17**, 184.

Sneath, P. H. A. (1957b) The application of computers to taxonomy. *J. gen. Microbiol.*, **17**, 201.

Sneath, P. H. A. (1967) Numerical taxonomy: steps in preparing taxonomic data for the computer. *Class. Soc. Bull.*, **1**, 14.

Sneath, P. H. A. (1968) Vigour and pattern in taxonomy. *J. gen. Microbiol.*, **54**, 1.

Sneath, P. H. A. (1971) Numerical taxonomy: criticisms and critiques. *Biol. J. Linn. Soc.*, **3**, 147.

Sneath, P. H. A. (1972) Computer taxonomy. *In*: *Methods in Microbiology*, J. R. Norris & D. W. Ribbons (Eds), Volume 7A, pp. 29–98. London: Academic Press.

Sneath, P. H. A. (1974a) Phylogeny of microorganisms. In Carlile and Skehel, 1974 (*q.v.*) p. 1.

Sneath, P. H. A. (1974b) Test reproducibility in relation to identification. *Int. J. Systemat. Bacteriol.*, **24**, 508.

Sneath, P. H. A. & Collins, V. (1974) A study in test reproducibility between laboratories: Report of a Pseudomanas Working Party. *Antonie van Leeuwenhoek*, **40**, 481.

Sneath, P. H. A. & Johnson, R. (1972) The influence on numerical taxonomic similarities of errors in microbiological tests. *J. gen. Microbiol.*, **72**, 377.

Sneath, P. H. A. & Skerman, V. B. D. (1966) A list of type and reference strains of bacteria. *Internat. J. Systemat. Bacteriol.*, **16**, 1.

Sneath, P. H. A. & Sokal, R. R. (1973) *Numerical Taxonomy. The Principles and Practice of Numerical Classification*. 573 pp. San Francisco: W. H. Freeman.

Snell, J. J. S. & Lapage, S. P. (1973) Carbon source utilization tests as an aid to the classification of non-fermenting Gram-negative bacteria. *J. gen. Microbiol.*, **74**, 9.

Sokal, R. R. & Sneath, P. H. A. (1963) *Principles of Numerical Taxonomy*. pp. 359. San Francisco: W. H. Freeman.

Stevens, M. (1969) Development and use of multi-inoculation test methods for a taxonomic study. *J. med. Lab. Tech.*, **26**, 253.

Subak-Sharpe, J. H., Elton, R. A. & Russell, G. J. (1974) Evolutionary implications of doublet analysis. In Carlile and Skehel, 1974 (*q.v.*). p. 131.

Subcommittee on the Taxonomy of Mycoplasmatales. (1972) Proposal for minimal standards for descriptions of new species of the order Mycoplasmatales. *Internat. J. Systemat. Bacteriol.*, **22**, 184.

Sueoka, N. (1961) Variation and heterogeneity of base composition of deoxyribonucleic acids: a compilation of old and new data. *J. Molec. Biol.*, **3**, 31.

Swartz, M. N., Trautner, T. A. & Kornberg, A. (1962) Enzymatic synthesis of deoxyribonucleic acid. XI. Further studies on nearest-neighbour base sequences in deoxyribonucleic acid. *J. biol. Chem.*, **237**, 1961.

Trotman, R. E. (1972) The application of automatic methods to diagnostic bacteriology. *Bio-med. Eng.*, **7**, 122.

Tsukamura, M. & Mizuno, S. (1968) "Hypothetical mean organisms" of Mycobacteria. A study of classification of Mycobacteria. *Jap. J. Microbiol.*, **12**, 371.

Ur, A. & Brown, D. F. J. (1975) Impedence monitoring of bacterial activity. *J. med. Microbiol.*, **8**, 19.

Walker, P. M. B. & McLaren, A. (1965) Fractionation of mouse deoxyribonucleic acid on hydroxyapatite. *Nature, Lond.*, **208**, 1175.

Wayne, L. G., Dietz, T. M., Gernez-Rieux, C., Jenkins, P. A., Käppler, W., Kubica, G. P., Kwapinski, J. B. G., Meissner, G., Pattyn, S. R., Runyon, E. H., Schröder, K. H., Silcox, V. A., Tacquet, A., Tsukamura, M. & Wolinsky, E. (1971) A co-oprative numerical analysis of scotochromogenic slowly growing mycobacteria. *J. gen. Microbiol.* **66**, 255.

Wayne, L. G., Engbaek, H. C., Engel, H. W. B., Froman, S., Gross, W., Hawkins, J., Käppler, W., Karlson, A. G., Kleeberg, H. H., Krasnow, I., Kubica, G. P., McDurmont, C., Nel, E. E., Pattyn, S. R., Schröder, K. H., Showalter, S., Tarnok, I., Tsukamura, M., Vergmann, B. & Wolinsky, E. (1974) Highly reproducible techniques for use in systematic bacteriology in the genus *Mycobacterium*: tests for pigment, urease, resistance to sodium chloride, hydrolysis of Tween 80, and $\beta$-galactosidase. *Int. J. Systemat. Bacteriol.*, **24**, 412.

Weed, L. L. (1963) Effects of copper on *Bacillus subtilis*. *J. Bacteriol.*, **85**, 1003.

Wildy, P. (1971) *Classification and nomenclature of viruses; first report of the International Committee on Nomenclature of Viruses.* 81 pp. Basel: Karger.

Willcox, W. R., Lapage, S. P., Bascomb, S. & Curtis, M. A. (1973) Identification of bacteria by computer: theory and programming. *J. gen. Microbiol.*, **77**, 317.

Wilson, G. D. & Miles, A. (1975) *Topley and Wilson's Principles of Bacteriology, Virology and Immunity*, 6th ed. 2706 pp. (2 volumes). London: Edward Arnold.

World Federation for Culture Collections Statutes (1972) *Int. J. Systemat. Bacteriol.*, **22**, 407.

Zvirbulis, E. and Hatt, H. D. (1969a) Status of names of bacterial taxa not evaluated in Index Bergeyana, Addendum II (1966). *Acetobacter* to *Butyrivibrio* (1966). *Int. J. Systemat. Bacteriol.*, **19**, 57.

Zvirbulis, E. & Hatt, H. D. (1969b) Status of names of bacterial taxa not evaluated in Index Bergeyana, Addendum II. *Achromobacter* to *Lactobacterium*. *Int. J. Systemat. Bacteriol.*, **19**, 309.

# 5 | Bacteroides Infections: Diagnosis and Treatment

## DONALD A. LEIGH

## Introduction

The significance and clinical importance of infections caused by anaerobic Gram-negative non-sporing bacilli has been known for many years but there are still many difficulties associated with their recognition.

The first report of bacteria living in the absence of oxygen was made by Pasteur between the years 1861 to 1863 investigating the production of butyric acid and the role of bacteria in putrefaction. He found a large bacillus which failed to grow in air and introduced the term anaerobe. In 1884 Loeffler observed similar bacteria in calf diphtheria and in 1891 Lewy recognized a small bacillus present in an abscess associated with gas formation which did not grow on culture. A few years later the first successful isolations from clinical material of Gram-negative bacilli belonging to the genus Bacteroidaciae were carried out by Vincent (1896), Veillon and Zuber (1898) and Hallé (1898), and subsequently these organisms were found to exist in various parts of the body as the normal bacterial flora (Tunnicliff 1906, Slanetz & Rettger 1933, Eggerth & Gagnon 1933).

In the ensuing fifty years many reports of the occurrence of anaerobic Gram-negative non-sporing bacilli in numerous sites of infection have been published and this versatility in manifestation and pathogenesis of disease has led to a great interest in the techniques of isolation and identification of these organisms. However, there is still little agreement in the field of taxonomy and many laboratories over the world fail to isolate and recognize these bacteria. In consequence there is little reliable information regarding the true incidence of the bacteroides group of organisms in all types of infection and even the most reliable of studies, especially those published before the mid 1960s, may provide a gross underestimation of their frequency.

The awareness of the possibility of an anaerobic infection by the clinician is the first step in a chain of events that leads to successful isolation of the bacteria in the laboratory and appropriate chemotherapy for the patient. Failure to collect the clinical specimen with the precautions necessary for the preservation of fastidious anaerobic bacteria is the principal factor in the low recorded incidences of these organisms as pathogens.

Because of the problems associated with identification of the bacteroides group of bacteria and the lack of agreement in their classification, there is great confusion in the literature between the various species. In this review of infections due to anaerobic Gram-negative non-sporing bacilli, emphasis will be made on the organisms belonging to the species Bacteroides, and in particular the subspecies fragilis which is isolated most commonly in the routine clinical laboratory. Reference to other members of the species will be made, especially in the sections on classification and taxonomy and identification, but in general and except where another subspecies is specifically named, the term bacteroides will refer to *Bacteroides fragilis*.

## Classification and Taxonomy

One of the main difficulties associated with the identification of non-sporing anaerobic Gram-negative bacilli has been the lack of a suitable classification system. The generic name Bacteroides was first used by Castellani and Chalmers (1919) and Eggerth and Gagnon (1933) attempted a further scheme based on organisms isolated from the gastrointestinal tract. They divided the 118 Gram-negative bacteria into 18 subspecies on the basis of morphology and carbohydrate fermentation, and it is likely that all these strains belonged to the fragilis group (Spiers, 1971). Slanetz and Rettger (1933) devised a system of 'grouping' of organisms isolated mainly from the mouth and Finegold et al. (1967) were later to follow their example. Prevot (1938, 1948, 1957, 1966) extensively studied anaerobic bacteria using highly specialised techniques not adaptable to the routine laboratory, and produced his own classification, but unfortunately used his own generic names, e.g., Ristella for Bacteroides, which were different from others in common usage at that time. Bergey et al. (1923), and a classification described by the later editors of Bergey's manual (Breed et al., 1957), described 30 species of bacteroides, only 3 of which were proven pathogens, the remainder being commensals of the alimentary tract. Although the initial classifications of anaerobic bacteria were based mainly on morphology, other characteristics of these organisms were gradually introduced such as the effect of dyes and bile on growth, and fatty acid production, particularly butyric acid (Buttiaux et al., 1962, 1966, 1969, Beerens et al., 1963), serological characteristics (Sonnenwirth, 1960) and the DNA base ratio (Sebald 1962). Suzuki et al. (1966) proposed glutamic acid decarboxylation and threonine de-amination as new characters. Other personal classifications were proposed by Rosebury (1962), Quinto (1964), Loesche and Gibbons (1965) and Werner (1965) which led to further difficulties.

The first universal agreement regarding classification and taxonomy occurred in 1967 at an International Sub-committee at Lille where major and minor characteristics were defined. The major characteristics were morphology, motility, final pH in glucose medium, threonine dehydration and butyric acid production. Promotion of growth by bile and inhibition by dyes, antibiotic sensitivities, glutamic acid decarboxylation, DNA and deoxyribonuclease, were classed as minor characters (Beerens, 1970).

Despite this international agreement, many exceptions and inter-mediate organisms have been described. Ninomiya et al., (1972) agreed that the production of butyric acid was a primary factor in the generic differentiation of anaerobic Gram-negative bacilli. However, this test requires expensive equipment and prolonged incubation (Cato et al., 1970).

There has been major disagreement between clinical and research laboratories, Finegold et al., (1967) disagreed with the classification used in Bergey's manual in view of the difficulty in carrying out complicated tests in routine laboratories and they preferred a system of grouping the organisms according to cell morphology, biochemical reaction and antibiotic susceptibility. They described 5 main groups, *B. fragilis*, *B. melaninogenicus* group, *B. oralis* group, Sphaerophorus and Fusobacterium. The strains in the *B. fragilis* group having rounded ends, showing pleomorphism, stimulation of growth in 10 per cent bile, the end pH in fermentation of glucose being less than 5·4, and resistance to penicillin but sensitivity to erythromycin.

Barnes and Goldberg (1968) carried out a computer analysis of 72 named strains and described 4 phena, all members of bacteroides apart from *B. melaninogenicus* composing phenon 2. They found the most useful differentiating tests to be cell morphology, terminal pH in glucose broth, formation of formic, acetic, propionic and butyric acids, production of propionic acid from threonine, growth stimulation by bile, the effect of various inhibitors, and earlier they had described the effect of antibiotics (Barnes *et al.*, 1966). There were however several anomalies and incorrectly named strains from the type culture collections examined, stressing the need for further investigation.

It would appear that despite the thorough and extensive studies in the past, the criteria and methods described initially by Finegold et al. (1967) and later by Finegold and Miller (1968); Shimada et al., (1970); Finegold, Sutter et al. (1970a), and Sutter and Finegold (1971), are most appropriate for the clinical microbiological laboratory. As a practical procedure it should be possible to combine the diagnostic and therapeutic antibiotics in a multiple disc (Multodisk) or ring (Masting), to allow easy identification of anaerobic Gram-negative bacilli in the laboratory and at the same time provide, without further delay, a range of antibiotic sensitivities for appropriate chemotherapy of infections.

## Normal Habitat

Anaerobic bacteria are widely distributed throughout the body and are commensals of the mouth, lower intestine and female genital tract. The presence of bacteroides in normal mouths was demonstrated by Tunnicliff (1906), Burdon (1928) and Slanetz and Rettger (1933), and has been confirmed by many later dental workers, (Gibbons et al., 1963). Eggerth and Gagnon (1933) first reported the presence of anaerobes in the faeces and found that bacteroides were present in 90 per cent of specimens. Weiss et al. (1937) suggested that appropriate methods of culture were necessary for accurate isolation and that if these were used bacteroides greatly outnumbered *E. coli* in the normal faeces. The main difficulty in the quantitative assessment of the bacteria in faeces is the ability to cultivate all the anaerobic bacteria present which might be killed by exposure to oxygen (Hine & Berry, 1937). Haenel (1961) carried out extensive studies and suggested that the largest group of bacteria were anaerobic lactobacilli with bacteroides the second largest group. However van Houte and Gibbons (1966) claimed bacteroides to be the commonest group of organisms and Drasar (1967) in a study using strict methods of anaerobiosis confirmed this, finding the count of bacteroides to be $10^{11}$ organism per gram of faeces. Anaerobic lactobacilli were equally common and all other species including enterobacteria were present in counts of $10^{6.4}$ or less. Drasar et al., (1966) previously had reported that 98 per cent of the viable bacteria in faecal specimens were bacteroides and less than 0·1 per cent were *E. coli*.

The importance of the normal flora of the intestinal tract as a signi-

**Table 5.1.** *Differential characteristics of the main species of Gram-negative anaerobic bacteria*

| | Bacteroides | Sphaerophorus | Fusobacterium |
|---|---|---|---|
| Microscopic Morphology (Fluid Thioglycollate medium) | Moderate pleomorphism | Pleomorphic with swellings & round bodies | Slender. Tapered ends. |
| Colonial morphology (Blood agar) | Grey white smooth entire (*B. melaninogenicus*, brown black) | Translucent colourless, greening, flat 'fried egg' appearance | Yellow green opaque, greening |
| Growth characteristics in broth (Thioglycollate) | Dispersed | Dispersed | Granular |
| Motility | Variable | Nonmotile | Nonmotile |
| Final pH in glucose broth | 5·5 (*B. melaninogenicus* 5·2–7·0) | 5·6–6·5 | 6·0–6·9 |
| Gas from glucose | Variable | + | Variable |
| Proprionic acid from Threonine | − | + | + |
| n-butyric acid production from glucose fermentation | − (*B. melaninogenicus* +) | + | + |
| Effect of 10% bile | No inhibition, may stimulate (*B. oralis* – inhibition) | No inhibition | Inhibition |
| Effect of dyes (Brilliant green 1:1000·000) (Victoria blue 1:140·000) | Sensitive | Resistant | Resistant |
| Fermentation of carbohydrates (other than glucose) | Very active | Active | Inactive |
| Glutamic acid decarboxylation | + | − | − |
| DNA base ratio GC% | >41% | <35% | <35% |
| Deoxyribonuclease | + | Variable | Variable |

ficant factor in natural resistance to infection was described by Nissle (1916) and other workers (Fredericq & Levine 1947) had shown that the production of colicines by some bacteria influences the growth of other species. The susceptibility to other bacterial infections following the suppression or elimination of the normal flora by antibiotic therapy has been well described, Freter (1955), and Hentges (1970) has shown in mice that the growth of shigella is inhibited by the presence of established cultures of bacteroides, but not if the two organisms grow together. Earlier Miller and Bohnhoff (1963) had correlated the disappearance of bacteroides from the faecal flora with increased susceptibility to salmonella infections. It has been shown that the numbers of bacteroides in the jejunal juice is increased in ill patients (Draser et al., 1966), although Mallory et al. (1973) sampling from five sites in the upper intestine could not detect any increase in patients undergoing resection of the lower bowel.

The presence of bacteroides in the vagina was reported by Burdon (1928) who was able to isolate *B. melaninogenicus* from 28 out of 35 normal women. Hite et al. (1947), found anaerobes in 70 per cent of cultures from the uterine cavity of postpartum women, although in antenatal women these organisms were commonly absent. Lukasik (1963) was only able to isolate *B. fragilis* rarely from the cervix and fallopian tubes, but Gorbach et al. (1973), found potentially pathogenic anaerobic bacteria in 70 per cent of cervical cultures from healthy women.

Wierdsma and Clayton (1964) found that antibiotics given to postpartum women did alter the bacterial flora but there was no change in the incidence of bacteroides which however was uncommon. The incidence of bacteroides in vaginal swabs taken from 246 pre-operative gynaecological patients was reported as 8·6 per cent by Neary et al. (1973), who considered them important pathogens which were more frequently isolated in the first half of the menstrual cycle. In 500 women attending a family planning clinic, the incidence of bacteroides isolation was 4·6 per cent and in 200 patients attending gynaecological outpatients 5 per cent (Leigh et al., 1976). These workers, however, did not find any influence of contraceptive measures or the stage of the menstrual cycle on the incidence of bacteroides and other anaerobic bacteria isolated from vaginal swabs.

## Methods of Isolation

The isolation of bacteroides from clinical material is dependent on three major factors, the collection of the specimen, the method of transportation and delivery to the laboratory, and the techniques of culture. All three factors are interdependent and equal in importance and therefore successful isolation requires the active co-operation of the clinician and the laboratory staff.

## Collection of the Specimen

In common with all bacteriological specimens, the laboratory can only isolate bacteria which are present in the material and remain viable during the interval before culture is carried out. Unlike aerobic bacteria which can remain viable in exudate, on the skin or dressings, anaerobes will probably die if the specimen is not collected from the primary focus of infection. In addition the relative numbers of bacteria in a superficial exudate from a lesion may not be representative of the basic infection, and it may be necessary to carry out dressing of the surface of the wound before collecting the specimen. Where anaerobic bacteria are suspected, it is essential that the collection of the specimen is not left to a junior nurse who does not understand the precautions that should be taken. In most instances of anaerobic infection, collection of a volume of pus or exudate will result in a higher isolation rate than a swab even though this is immediately placed in a transport medium.

There are many good reasons for the laboratory providing a collection service, as this will result in a greater incidence of isolation and in terms of the hospital will be more cost effective, allowing appropriate antibiotic therapy to be given more rapidly to the patient.

## Transportation of the Specimen to the Laboratory

In practice the commonest cause of failure to isolate anaerobic bacteria in a routine laboratory is the lack of an adequate transportation system. It is useless to try and isolate anaerobes from a swab which has been drying and has been in contact with oxygen for an hour or more. Percival and Roberts (1972) showed that bedside culture of wound swabs resulted in a large increase in the isolation rate of anaerobes. It is essential to use a form of culture medium that provides adequate anaerobiosis and prevents overgrowth of aerobic bacteria which may be able to multiply in the interval before laboratory examination. There are two main forms of transporting specimens, first in a transport medium, allowing culture techniques to be carried out in the laboratory, and second where culture is carried out at the bedside into special media. The most appropriate transport media are either Stuarts

F

transport medium, which is a minimal nutrient agar providing anaerobic conditions and which will maintain most fastidious bacteria for a short period of time, or Robertson's meat broth or Brewer's medium, which can be used as a culture medium and subculture can be carried out at the appropriate times. For the routine laboratory these methods of transportation are the most acceptable as the technicians are not involved in the collection of the specimen. Where facilities and staff are available, there is no doubt that the bedside innoculation–culture techniques result in a greater isolation rate of anaerobes. Many varieties of bottles and tubes have been described, the first being the pre-reduced anaerobically sterilized (PRAS) roll tubes introduced by Hungate (1950) and modified by Moore (1966) and described by other workers (Attebery & Finegold, 1969; Anaerobic laboratory, VPI manual 1970), which can contain agar or broth, and be innoculated at the bedside. Davis et al. (1973), described the similar use of a flat bottle. All these methods require the innoculation of a liquid sample, but many infections may contain thick pus which is not easily aspirated. Whatever method of specimen collection is employed, it is important that the container reaches the laboratory as quickly as possible to prevent overgrowth due to more actively growing aerobic bacteria.

## *Laboratory culture techniques*

In spite of precautions that may be taken in the collection and transportation of specimens containing anaerobic bacteria, it is important that special selective procedures are used in the laboratory. Bacteroides, besides requiring strict anaerobic conditions, usually grow better in the presence of carbon dioxide (Watt, 1973), but they grow much more slowly than the majority of aerobic bacteria. The anaerobic conditions can be created by the use of special anaerobic cabinets, jars exhausted of oxygen and replaced by nitrogen, hydrogen or a hydrogen–carbon dioxide mixture, or a liquid medium containing chemicals with strong reducing properties. Various designs of anaerobic cabinets have been produced from store boxes (Socransky et al., 1959; Rosebury & Reynolds, 1964) to larger cabinets (Drasar, 1967; Killgore et al., 1973). The conventional jar for isolating anaerobes was originally described by Brown (1921, 1922) and later modified by Brewer (1939) and Evans et al. (1948); these jars incorporated electrically heated catalysts, but were improved by the use of a cold catalyst (Heller, 1954; Khairat, 1964), however it was still necessary to use a cumbersome hydrogen generating system or cylinders of compressed gases which involved explosive risks. The introduction of a borohydride system (Brewer et al., 1955) led to the production of a combined hydrogen carbon dioxide system in a foil envelope available as the Gaspak (BBL, Maryland) and allowed specially designed light weight plastic jars to be used for anaerobic culture. This Gaspak method has been found to be more reliable and reproducible than other methods (Collee et al., 1972), although the incorporation of carbon dioxide has led to minor difficulties in the reporting of antibiotic sensitivities (Ingham et al., 1970). Thioglycollate broth has been most commonly used as the liquid medium. In the routine laboratory, however, it is necessary to provide an assessment of the relative numbers of different species of bacteria present in a specimen and solid agar plates are used in conjunction with a liquid medium, either thioglycollate or Robertson's meat broth, which allow for subculture after a few days. Anaerobic conditions can be met by the simultaneous growth of aerobic bacteria such as *E. coli* in the medium which reduce the oxygen supplies through catalase activity, but this is not applicable to the routine laboratory. The main difficulty in reading culture plates is to detect the presence of anaerobic bacteria amongst large numbers of rapidly multiplying aerobic organisms. The isolation of all anaerobic organisms in a specimen may require a large number of agar plates. Drasar (1967) describes the use of 14 such media to identify all anaerobes to be found in the faeces, but this is not necessary for most specimens from clinical infections. To this end most laboratories use special selective media which will reduce or inhibit the growth of aerobic bacteria. This can be achieved by the use of chemicals, for example phenylethyl alcohol (Dowell et al., 1964) or antibiotics incorporated into the agar. Bacteroides and other anaerobes are usually resistant to the aminoglycoside group of antibiotics, and neomycin and paramomycin are most commonly used, although kanamycin is equally appropriate, the concentration being usually 100 µg per ml which inhibits the growth of most naturally occurring aerobic bacteria, although *Pseudomonas aeruginosa* and hospital strains with acquired resistance may grow through. For the identification of other anaerobic species the use of rifampicin 50 µg/ml will inhibit bacteroides and allow the isolation of Fusobacterium (Sutter et al., 1971). The antibiotics are usually incorporated into blood agar plates but Brucella agar or brain heart infusion agar can also be used. For the isolation of some anaerobes it is also necessary to add haemin (5 mg/l) and menadione or Vitamin K (0·5 mg/l) (Gibbons & Mac-Donald, 1960).

Although the use of complex laboratory methods does result in the isolation of a higher percentage of anaerobic bacteria under experimental conditions, this is probably not true of clinical specimens (Rosenblatt et al., 1973) where the cultural techniques should be comprehensive but simple.

It is necessary to incubate the agar plates for several days as while *B. fragilis* has usually grown to produce macroscopic colonies by 24 or 48 h, other species of anaerobes may take up to 3 weeks to grow, Therefore it is an advantage to use a liquid medium in addition to solid agar for primary culture and to subculture onto selective media so that the anaerobes can be isolated in pure growth.

## Identification

The identification of anaerobic bacteria is carried out by the usual methods of colonial and microscopic morphology, biochemical characteristics and antibiotic susceptibility patterns. In the routine laboratory, the primary identification is the growth of an organism on the selective medium. Special diagnostic agars have been used which incorporate various dyes, such as China Blue (van de Wiel–Korstanje & Winkler, 1970) and sodium azide and ethyl violet (Post et al., 1967) which result in typical coloured colonies identifying bacteroides, sphaerophorus and fusobacterium. Unfortunately these measures are not really applicable to the routine laboratory. The microscopic appearances of anaerobic bacilli are very variable, but may be useful when regular forms are seen. The biochemical reactions can be helpful in the research laboratory where the end pH of glucose fermentation, threonine degradation and production of various acids can sometimes produce a definitive result, however most of these tests take too long to be of value to the clinical laboratory. The sensitivity to bile and sodium desoxycholate (Shimada et al., 1970) can be used. Fluorescent antibody techniques have been applied successfully to the identification of anaerobes (Griffin, 1970). The antigenic structure of bacteroides has ·not been completely worked out but Romond et al. (1972), described 6 major and 6 minor subgroups denoted by the letters A to E and a to e respectively. These authors reported that only strains with antigen E and e are pathogenic. Specific antisera have been produced to the main species of Bacteroides (de la Cruz & Cuadra 1969) but little study has been carried out in this field.

The susceptibility to antibiotics is probably the best aid to the identification and characterization of anaerobic Gram-negative bacilli in the routine clinical laboratory (Finegold et al., 1967). Six antibiotic discs are used, penicillin G (2 units), erythromycin (60 g), kanamycin (1000 µg), colistin (10 µg), rifampicin (15 µg), and vancomycin (5 µg), and the main species of the Bacteroidaciae can be distinguished (Table

5.2, p 126). It is doubtful whether any further identification is of value other than under research conditions.

## Pathogenicity

Whilst bacteroides are common commensals of many parts of the body there is no doubt of their pathogenicity and ability to produce devastating disease. One of the important defences of the body against anaerobic infection is the normal EH ( +120 mV). Alteration of this potential through impaired blood supply, tissue necrosis or growth of aerobic bacteria, allows anaerobic bacteria to spread throughout various tissues and in addition impairs the phagocytic and bactericidal activity of white cells.

The virulence of anaerobic Gram-negative bacilli and their ability to invade tissues and become localized in abscesses (Chandler & Breaks, 1941) results from toxin production. The endotoxin (Lipopoly saccharide, LPS) produced by bacteroides is similar to that found with aerobic infection, but it does not contain heptose as the basal sugar or KDO as a bridge between the polysaccharide and lipid (Hofstad & Kristoffersen 1970). In bacteroides an unknown amino compound is present, which can be detected by Limulus assay techniques (Sonnenwirth et al., 1972). Although Garrod (1955) suspected penicillinase production in 2 strains of *B. fragilis* and a few workers have found β-lactamase activity, Holt and Stewart (1964) found no β-lactamase present in 8 strains and this was confirmed by Okubadejo et al. (1974) even after disintegration of the cells. Pinkus et al. (1968), however suggested that the β-lactamase was probably intracellular as the activity was greater when the bacterial cells were disrupted and Anderson and Sykes (1973) described an intracellular β-lactamase more active against the cephalosporins than penicillins which was not inducible or transferable.

An important toxin produced particularly by *B. fragilis* is heparinase, first reported by Gesner and Jenkin (1961) which is responsible for the thrombophlebitis, characteristic of all bacteroides infections (Reid et al., 1945). This lesion may predispose to metastatic abscesses all over the body, particularly in the liver, lungs, and central nervous system.

## Diagnosis of Infection

There are many clinical clues to the possibility of an anaerobic infection. All anaerobic infections can produce a foul smelling discharge in contrast to many aerobic infections which are characterized by a lack of odour. Many infections attributed to *E. coli* because of a 'typical smell' are due to anaerobes which are not isolated by the laboratory, and the pus associated with *E. coli* infections has no typical odour. Where infection occurs near to a mucosal surface, e.g., the intestine, genital tract or mouth, or faecal contamination has occurred following perforation of the intestine, there is a high probability of an anaerobic infection. However, in many cases the possibility of contamination rather than infection by anaerobic bacteria must be considered.

The presence of septic thrombophlebitis is a primary diagnostic feature and this may be associated with jaundice and pylephlebitis. Unless adequate chemotherapy is given metastatic abscesses in many parts of the body are commonly seen in anaerobic infections. Proven endocarditis with persistent negative blood cultures or bacteraemia associated with jaundice should suggest an anaerobic etiology.

In clinical practice the most common indication of the presence of anaerobes is the failure of an infection to respond to antibiotic therapy, particularly the aminoglycosides, or the development of infection during the course of therapy with this group of chemotherapeutic agents. Unlike all other members of the enterobacteria, bacteroides are highly resistant to this group of antibiotics.

Norris (1901) reported the frequency in which bacteroides were associated with anaerobic streptococci in infective lesions and other workers have reported incidences of approximately 10 per cent where mixed cultures of anaerobes were present (Zabransky, 1970; Leigh, 1974).

The relationship of specific circulating antibodies to bacteroides in the blood has not been extensively studied, Vinke and Borghans (1963) showed an increase in antibody to titres of 1/640 in patients with bacteroides infections. Brown and Lee (1974), using radio-immunoassay techniques showed that the concentration of specific IgG IgA and IgM was considerably higher in patients with Crohn's disease and ulcerative colitis than in controls or sprue where the increases were lower. However, the antibody concentration was not related to bacterial population. It is not known what percentage of the general population have increased antibodies and what level of antibodies is indicative of active infection as against past infection. This technique, however, could provide a useful tool in the diagnosis of chronic disease of the gastrointestinal tract.

In the laboratory certain clues may be present which will aid in the diagnosis of possible anaerobic infections. On a Gram film many anaerobes have a characteristic morphology and with experience the presence of bacteroides can be detected. The failure to grow bacteria, when seen in the Gram film of the original sample, or the presence of bacterial growth on special selective agar which may contain inhibitory or additional growth factors, are primary indications of the presence of anaerobes. Frequently the foul odour present in culture media, solid agar or broth, is an obvious indication especially when an anaerobe is present in mixed culture. In the laboratory many anaerobic bacteria are not detected because of a failure to proceed with further diagnostic identification procedures. It is not sufficient to assume that Gram-negative bacilli isolated from the anaerobic culture plates are the same species as seen on aerobic plates, even though there may be colonial and cellular similarity. Bacterial growth on anaerobic culture media should be assumed to be different from that on aerobic culture media until proven by laboratory procedures to be identical.

## Isolation Rate

There is little doubt that until adequate methods of specimen collection and laboratory culture are universally used, the true incidence of infections will remain unknown. The reported incidences of isolation in many instances are so low as to be unbelievable for bacteria which heavily colonize many areas of the body, and it is still common to find published reports of small numbers of cases. However, because anaerobic bacteria are more likely to be isolated from infections in certain parts of the body, it may always be impossible to report a true isolation rate and compare results between different laboratories and hospitals, as the incidence will depend on the type of patient studied.

Although there were many reports of bacteroides infections in the late nineteenth and early twentieth centuries, there were no quoted incidences in hospital patients, and numbers varied greatly. Dack (1940) saw 105 patients in 4 years, Beigelman and Rantz (1949) 47 cases in 8 years, McVay and Sprunt (1952) 35 cases from widely differing sources in 5 years, and Lodenkamper and Steinen (1955) 690 in 8 years. Bornstein et al. (1964) noted an increase in the incidence of anaerobic infections from 0·9 per cent in 1960 to 2·0 per cent in 1963, and Smith et al. (1968) found the overall percentage isolations from clinical material was 2·66 per cent; Saksena et al. (1968) in a 2 year period grew bacteroides from 1·9 per cent of specimens. In laboratory studies however, Stokes (1958) isolated anaerobes in 496 of over 5000 specimens, Zabransky (1970) found bacteroides in 26 per cent of specimens, and Mitchell (1973) in 1067 of over 12000 specimens. Hoffmann and Gierhake (1969) using improved techniques, found an increased isolation in wound swabs from 5 per cent in 1964 to 33 per cent in 1967. Leigh (1974) showed that the incidence after intestinal surgery was increased from 13 per cent to over 80 per cent using similar techniques, and Percival and Roberts (1972) through culture at the bedside increased the incidence from 30–70 per cent. Finegold and Rosenblatt (1973) suggested that *B. fragilis* was responsible for 10 per cent of aspiration pneumonias, and Rotheram and Schick (1969) suggested that bacteroides were the commonest organisms in septic abortion.

In special specimens such as blood cultures, the isolation rate of bacteroides has varied between approximately 1·0 per cent (McHenry et al. 1961; Mitchell & Simpson 1973) and 6·0 per cent (Hermans & Washington, 1970; Kagnoff et al., 1972).

Bacteroides are usually isolated in mixed cultures with many other organisms but pure cultures are not uncommon. In general, infections of genitourinary origin are more likely to be associated with pure growth than those of intestinal origin (Okubadejo et al., 1973; Leigh, 1974) and the important factor in infections from the gastrointestinal tract is the presence of perforation when mixed cultures are twice as common (Leigh et al., 1974).

## Gastrointestinal Tract Infections

The first report of bacteroides in intestinal infections was by Veillon and Zuber (1898) who isolated the organism from 25 cases of gangrenous appendicitis. Weinberg et al. (1928) and Schmitz (1930) noted the association of bacteroides with the normal and diseased appendix and Lemierre (1936) stressed the frequency with which bacteroides bacteraemia resulted from appendicitis. Gillespie and Guy (1956) isolated bacteroides from 11 of 13 cases of acute appendicitis and 67 out of a total of 111 patients with various suppurative lesions of the abdomen. Altemeier (1938) noted that 96 of 100 cases of acute appendicitis were associated with bacteroides. In a review of published cases, Gunn (1956) found 26 of 148 cases of bacteraemia had disease of the gastrointestinal tract, and only 5 had acute appendicitis although a further 9 had peritonitis of unknown etiology which was treated conservatively. The frequency of appendicitis reported as the primary factor in the development of post operative infection is probably only related to the commonness of the condition. Bornstein et al. (1964) found that a quarter of their cases involved the appendix, and Okubadejo et al. (1974) reported that 41 of 112 patients had had appendicectomy or an appendix abscess. Mitchell (1973) reported 80 isolations in 256 cases. In a series of 200 bacteroides infections, Leigh (1974) found that 75 (27 per cent) were directly related to the appendix. Saksena et al. (1968) had previously only shown the appendix as the source of 11 of 112 cases, and Smith et al. (1968), in 1 per cent of isolations. These latter studies however provide a minority view, and it is likely that as bacteroides account for over 90 per cent of the faecal bacterial population, the incidence of isolation of these organisms should be high from specimens of intestinal origin where contamination of the peritoneal cavity or extra-peritoneal space has occurred. In swabs from the appendix fossa of patients undergoing appendicectomy, bacteroides were isolated in 78 per cent of cases where there was bacterial growth (Leigh et al., 1974). The majority of post-operative wound infections were caused by bacteroides and the presence of this organism on primary culture of the swab showed an increased susceptibility to post-operative complications. Heavy contamination by faecal material as occurred with perforation of the appendix was associated with post operative wound infection in 63 per cent of cases, unless appropriate chemotherapy was given. Although the isolation of bacteroides did not automatically imply infection, the presence of large numbers of this organism in the peritoneal cavity is a potential infective situation.

Dixon and Deuterman (1937) noted the association of malignancy of the large bowel and post-operative bacteroides infections, and Rubin and Boyd (1958) reported that intra-abdominal abscess and wound infections were related to diverticulitis and ulcerative colitis. Gunn (1956) included 8 cases of carcinoma of the rectum in his series of septicaemia, and more recent studies, Bornstein et al. (1964), and Leigh (1974), have confirmed this association, reporting 6 out of 20 and 11 out of 24 infections respectively following surgery for abdominal malignancy, the remaining cases following diverticulitis. In addition, conditions such as Crohn's disease and ischaemic ileitis and colitis are frequently complicated by post-operative wound infections.

In clinical practice the part played by bacteroides in infections associated with the gastrointestinal tract is the development of intra-abdominal abscesses or post-operative wound infections or abscesses. The significance of bacteroides in inflammatory conditions of the intestine, such as diverticulitis, colitis and appendicitis, is not known. While there is no doubt that these organisms, once allowed to penetrate the mucosal membrane through perforation or become lodged in an obstructed area, can give rise to secondary infection, whether they can be primary pathogens is unknown.

## Abscesses

Bacteroides are frequent infecting organisms in the many varieties of abscesses that occur in the human body. Although in most instances the organisms are derived from the oral cavity, gastrointestinal tract or urogenital tract and the abscesses occur in close proximity to these sites, the occurrence of bacteraemia can lead to metastatic abscess in any situation in the body, Gillespie and Guy (1956) were able to isolate anaerobic organisms from 60–80 per cent of their cases of abdominal abscesses, and a similar finding was reported by Altemeier et al. (1973). Cohen (1932) reported the frequency of isolation of bacteroides from lung abscesses, and these organisms have been reported in abscesses in the myocardium complicating infarction (Castleman & McNeely, 1970; Lewis 1973), in the thyroid gland (Hawbaker 1971) and in deep musculoskeletal sites (Nettles et al., 1969; Pearson & Harvey 1971) where they may have a chronicity of several years. Rein and Cosman (1971) described multiple abscesses due to bacteroides in the arm of a drug addict, and this was accompanied by necrotizing fasciitis. The brain as a site for bacteroides abscess has been reported by Ballantine and Shealy (1959) and Salibi (1964) and Schoolman et al. (1966), described a brain abscess occurring two months after peritonitis complicating hysterectomy. Pearson (1967) has reported bacteroides as a not uncommon cause of breast abscess.

In more recent studies, Leigh (1974) reported 51 cases of bacteroides abscess, 40 of which were derived from the gastrointestinal tract, and 9 from the genitourinary tract, and Baird (1973) found that 27 of 63 abscesses following surgery were related to the colon and rectum. The majority of abscesses relating to suppurative condition of the anorectal region are due to bacteroides and from this region deep seated abscesses of the gluteal region can occur (Lindell et al., 1973). Similarly a retropharyngeal or mediastinal abscess can occur from perforation of the oesophagus (Janecka & Rankow, 1971).

The isolation of many species of bacteria from abscess pus has been well studied by Altemeier et al. (1973), and Gorbach et al. (1974), and an average of 5 species including 2 aerobic and 3 anaerobic have been cultured. However, the relative pathogenicities of each species is not known and while there may be considerable interdependence between species, one providing a suitable environment for another, the importance of providing antibiotic therapy for all bacteria isolated is unknown. Many workers would argue for the major virulence of the anaerobes and frequently chemotherapy which excludes an antibiotic active against these organisms is unsuccessful.

## Liver Abscess and Biliary Tract Infections

Liver abscess is due primarily to 3 groups of bacteria, *B. fragilis*, anaerobic streptococci and *E. coli*, and these organisms may be mixed (Futch et al. 1973). Other anaerobes such as Fusobacterium, Clostridium and Actinomyces may also be found. Sabbaj et al. (1972) reviewed 25 cases of liver abscess and found 17 due to anaerobes. Anaerobic abscesses and infection are usually secondary to pylephlebitis rather than cholangitis (Finegold & Rosenblatt, 1973) but Butler and McCarthy (1969) in a report on 48 cases found that about one third presented after cholangitis, diverticulitis and appendicitis. Liver abscess has become rarer because of the improved treatment of appendicitis which was formerly a common cause. Futch et al. (1973), presented 5 cases of bacteroides liver abscess, 4 of which developed secondary to intra-peritoneal infection, and McLaughlin et al. (1973) reported a case complicating radiation enteritis.

Anaerobic bacteria are not commonly associated with infections of the biliary tract and the gall bladder, although bacteria have been isolated from approximately 50 per cent of gall bladders containing stones (Andrews & Henry, 1935; Fukunaga, 1973), and positive cultures were found in 70 per cent of gall bladder specimens at cholecystectomy (Edlund et al., 1958). Bacteroides have been rarely reported and strepto-cocci and clostridia are the commonest organisms. However, many studies have not used adequate anaerobic techniques. In a study at High Wycombe in 1972, *B. fragilis* was isolated from 12 per cent of specimens of bile from gall bladders containing stones.

## Female Genital Tract Infections

The first description of bacteroides in genital tract infections was made by Hallé (1898) who found the organisms in Bartholins abscesses. The frequency of bacteroides in abscesses related to the female pelvis was overlooked for many years, and anaerobic streptococci and *Clostridium welchii* were regarded as the most common pathogens (Schwarz & Dieckmann, 1927; Brown, 1930). However, while many of the early studies found a high percentage of sterile specimens, later workers have found anaerobes frequently in pelvic abscesses (Swenson et al., 1973, Thadepalli et al., 1973a) and in ovarian and adnexal abscesses, between 60 and 100 per cent are associated with these organisms (Ledger et al., 1968; Pearson & Anderson, 1970a). With abscesses of the Bartholin glands, although a small proportion are caused by *Neisseria gonorrhoea*, the majority are due to anaerobes including bacteroides (Parker &

Jones, 1966). Carter et al. (1951), found that of 133 patients with pyo-
metra 41 had mixed infections including bacteroides, and in a similar
study of 153 patients with abscesses of pelvic origin, they reported
that all of them were caused by bacteroides (Carter et al., 1953).

The importance of bacteroides in post partum infections was noted
by Harris and Brown (1927). Contamination of the uterine cavity can
occur with all deliveries, and Hite et al. (1947) were able to isolate
anaerobes from 70 per cent of post partum uterine cultures from normal
women. Pearson and Anderson (1970b) found that bacteroides infections occurred in one third of the infected cases following caesarian
section, and Ledger et al. (1971), found 11·8 per cent of infections in
obstetric patients due to these organisms, and commented that an
effort should be made to reduce the level of pathogens before operations.
Clark and Wiersma (1952) and McVay and Sprunt (1952) reported the
high incidence of severe illness in post partum women, and the frequency
of bacteraemia. Bacteroides infections are commonly associated with
premature rupture of the membranes and play a significant role in fatal
puerperal infections following caesarian section. Bacteraemia due to
bacteroides has a severe prognosis in pregnancy and is related to a high
incidence of perinatal death (Pearson & Anderson, 1967).

Severe bacteroides wound infections following episiotomy are common, and the association between post partum fever and the isolation
of bacteroides from a high vaginal swab is an indication of the frequency
of invasion by these organisms. Although the significance of bacteroides
isolation from the vagina in relation to uterine infection is often doubtful, specific chemotherapy such as clindamycin or lincomycin is usually
clinically successful. Interestingly, endometritis is less common after
vaginal deliveries than following caesarian section (Gorbach & Bartlett, 1974).

Septic abortion is commonly associated with anaerobic bacteria
(Pearson & Anderson, 1970b). Rotheram and Schick (1969), found that
where adequate cultures had been performed anaerobes were isolated
from 81 per cent of patients, whereas aerobes were only seen exclusively
with 6 patients. The high frequency has been confirmed by Thadepalli
et al. (1973a).

The major types of anaerobic infection of the female genital tract are
comprehensively tabled by Finegold and Rosenblatt (1973), but apart
from abscesses, the commonest type of bacteroides infection seen in a
district hospital is that complicating hysterectomy or corrective surgery
for prolapse and cystocele formation. Okubadejo et al. (1973), describe
15 of 29 infections, and Leigh (1974) 8 of 31 infections associated with
gynaecological diseases that followed hysterectomy. Hall et al. (1967).

noted that 5 per cent of infections associated with hysterectomies were localized in the vaginal cuff, and nearly all grew anaerobes and a rate of 76 per cent was found by Swenson et al. (1973).

In the clinical laboratory many routine vaginal swabs are found to grow *B. fragilis*. In most cases this indicates either inadequate collection of the specimen or colonization of the lower vagina, but in women with senile vaginitis there appears to be a true correlation between the inflammatory condition and the bacteria, as specific antibiotic therapy can cause relief of symptoms. However, senile vaginitis is primarily due to oestrogen deficiency and a complete cure cannot be achieved by the use of antibiotics alone.

## Bacteraemia

Hallé (1898) described the isolation of bacteroides from the blood of patients dying from suppurative conditions complicated by jaundice and bacteraemias have followed infections usually in the gastrointestinal, genital and pharyngeal areas (Lemierre, 1936). Although many reports of bacteraemia followed over the next thirty years (Tiessier et al., 1931; Thompson & Beaver, 1932; Brown et al., 1941; Reid et al., 1945; Beigelman and Rantz 1949), in 1956 Gunn when reporting 2 of his own cases of bacteraemia, was only able to find 162 other cases in the literature, and the commonest primary lesion was nasopharyngeal infection, particularly tonsilitis which was seen in 54 cases. McHenry et al. (1961), reviewing their records of bacteraemia from 1948–59 found only 1·4 per cent of patients had bacteroides bacteraemia, although the total incidence of anaerobes was 2·2 per cent. Crowley (1970) and Sinkovics and Smith (1970) have reported similar incidences of 1·6 per cent and 2·2 per cent respectively. However, other workers (Hermans & Washington, 1970; Kagnoff et al., 1972; Wilson et al., 1972; Sonnenwirth, 1974; Chow & Guze, 1975) have recently reported incidences of between 6 and 11 per cent.

There has therefore been a definite increase in the number of positive cultures and the most likely explanations are improved cultural methods and an increased awareness of the possibility of bacteroides infection. *B. fragilis* accounts for up to 90 per cent of the anaerobic isolates from blood cultures and this is expected in view of the frequency of probable infection sites in the gastrointestinal and genital tracts. The part played by chemotherapy in selecting anaerobic organisms is not known, but it is likely that this will follow the widespread use of the aminoglycosides group of antibiotics.

The 'shock syndrome' occurs in up to 35 per cent of patients (Gunn,

1956; Gelb & Seligman, 1970; Marcoux et al., 1970; Bodner et al., 1970; Wilson et al., 1972; Nobles, 1973), but there are distinguishing features in bacteroides bacteraemia which are diagnostically helpful, particularly the presence of jaundice and thrombophlebitis.

Bacteroides bacteraemia is closely associated with other conditions. Sinkovics and Smith (1970) and Kagnoff et al. (1972), all noted that there was a high incidence in patients with malignancy, although Gelb and Seligman (1970) found no correlation with the use of immunosuppressive therapy. Schoutens et al. (1973) and Ellner and Wasilauskas (1971), found bacteraemia more commonly in older patients where the mortality was also highest.

The majority of patients with bacteraemia have undergone intestinal surgery (Felner & Dowell, 1971; Wilson et al., 1972), unlike aerobic bacteraemia where the urinary tract is the commonest portal of entry (Brumfitt & Leigh, 1969). Bacteroides infection of the female genital tract also accompanies surgery or obstetric delivery and the incidence of bacteraemia is high. Pearson and Anderson (1970b) reported that 50 per cent of post-abortal women with severe bacteroides infections had bacteraemia.


## Endocarditis

Anaerobic bacteria are rarely implicated in endocarditis. Nastro and Finegold (1973) reported 37 cases found in the literature, 13 being due to *B. fragilis*. Most patients were over the age of 40 years, and the most likely source was the gastrointestinal tract (8 cases), and the respiratory tract (2 cases). Underlying cardiac disease is usually present in over 50 per cent of cases, but the incidence in large series of subacute bacterial endocarditis varied between 84 and 100 per cent (Cates & Christie, 1951; Friedberg et al., 1951). Nastro and Finegold (1973) suggested that anaerobic organisms are more likely to attack normal valves than the usual aerobic bacteria causing endocarditis. There are few reports of the incidence of anaerobes in endocarditis but Gorbach and Bartlett (1974) in an aggregate total from two large studies (Felner & Dowell, 1971; Weinstein & Rubin, 1973) reported an incidence of 3·8 per cent. The necessity for prolonged incubation of blood cultures for the isolation of bacteroides in patients with endocarditis has been stressed by Tumulty and McGehee (1948), McHenry et al. (1961), and it is essential to use special broth such as thioglycollate or Robertsons meat broth as anaerobic conditions may not be attained in nutrient or glucose broth.

## Urinary Tract Infections

The urethra, perineum and genital areas are colonised by both anaerobic and aerobic bacteria, and it is surprizing that anaerobes are not reported more frequently in urinary tract infections. Headington and Beyerlein (1966) in a large study were able to isolate anaerobic organisms from only 1·3 per cent of urines, and *B. fragilis* was only present in 7·3 per cent of these cases, but Kuklinca and Gavan (1969), using a non-quantitative technique with thioglycollate broth, reported anaerobes in 9 per cent of 107 specimens. Finegold et al. (1965), using the suprapubic puncture technique, found no anaerobes in 100 specimens, and in a review of 56 anaerobic infections of the urinary tract reported the commonest lesion to be periurethral abscess (27 patients). Leigh (1974) reported 5 cases of bacteroides infection of the urinary tract, two associated with prostatectomy, one each with pyonephrosis, carcinoma of the bladder, periurethral abscess and a vesicocolic fistula where the organisms originated primarily from the gastrointestinal tract. Bacteriological sampling of areas of the urinary tract without operative specimens is very difficult. Many cases of urethritis are associated with microaerophilic or anaerobic streptococci and it is likely that anaerobic bacilli are also fairly common pathogens in contrast to the findings in cystitis.

The bacteriology of prostatitis, particularly when chronic, and epididymo-orchitis is not well known and if more accurate sampling procedures were available, a significant number of these patients might be found to be suffering from anaerobic infections. The involvement of the perinephric space by anaerobic bacteria derived from the gastrointestinal tract has been reported and in obstructuve lesions of the kidney, such as pyonephrosis and renal tissue necrosis associated with calculi, anaerobic bacteria have been found (Finegold & Rosenblatt, 1973). The source of the bacteria in these cases is probably the gastrointestinal tract through haematogenous spread. Whilst the isolation of anaerobes from such lesions is evidence of their pathogenicity, the quantitative assessment of these organisms in urine, especially of asymptomatic patients, is difficult. Alling et al. (1973) found significant counts of anaerobic bacteria in 70 per cent of 44 incontinent geriatric patients, but these organisms were probably contaminants from the lower urinary tract. Considerably more information is required before the significance of anaerobic bacteria in the urinary tract is known.

## Bone and Joint Infections

Most anaerobic infections of bones and joints are haematogenous

(Hallé, 1898; Rist, 1898) and must spread from the mucous membranes. The most common site is reported to be the upper respiratory tract (Ziment et al., 1969). However, in the published cases of joint infections the commonest organism is *Sphaerophorus necrophorus* and *B. fragilis* has been reported less frequently. Kelly et al. (1970) found only 2 of 78 cases of pyogenic arthritis caused by anaerobes. Ziment et al. (1969), presented one case of *B. fragilis* arthritis of the knee following intra-articular steroid therapy, and Ament and Gaal (1967) described a case associated with only minimal trauma. Bacteroides arthritis has also been reported by Saksena et al. (1968), in adults and by Nelson and Koontz (1966) in children.

Bacteroides osteomyelitis has been reported by Chandler and Breaks (1941) in the hip following otitis media and in other bones by Beigelmann and Rantz (1949) and Gelb and Seligman (1970), and Statman and Spitzer (1962) described multiple sites of osteomyelitis in a patient with sickle cell disease. Pearson and Harvey (1971) report 60 infections in orthopaedic patients including 9 cases of osteomyelitis, and observed that 50 per cent had severe disorders such as diabetes, chronic alcoholism or drug addiction, confirming the view of Nettles et al. (1969), that these conditions may predispose to bacteroides infection.

Bacteroides infections may follow human or animal bites (Maier, 1937; Hubbert & Rosen, 1970), and osteomyelitis of the mandible and maxilla has been recorded sometimes following teeth extraction (Beerens & Tahon-Castel, 1965; Leake, 1972) which in the pre-antibiotic era carried a high mortality.

Post operative infections after bone surgery have been reported by Tynes and Frommeyer (1962) and Nettles et al. (1969).

## Respiratory Tract Infections

The significance of anaerobic bacteria in pulmonary infection derives from the continual aspiration of secretions from the upper respiratory tract, the aspiration of gastric contents and the frequency of metastatic spread of infections elsewhere in the body to the lungs via the blood stream. Whilst *B. fragilis* are common in lung abscesses and empyema the majority of pulmonary infections are associated with other anaerobes, such as fusobacterium, and Smith (1932) concluded that between 80 and 95 per cent of bronchiectasis, lung abscess and necrotizing pneumonia were caused by anaerobic bacteria. Anaerobic empyema was first described in 1899 by Rendu and Rist, and where routine anaerobic culture is carried out, anaerobes have been recovered from empyema fluid in between 8 and 76 per cent of cases (Bartlett & Fine-gold, 1972).

In lung abscesses, anaerobes are frequently encountered (Cohen, 1932), but may not be isolated without adequate precautions particularly in the collection of material for culture, such as transtracheal aspiration (Bartlett et al., 1974b) or percutaneous transthoracic aspiration (Beerens & Tahon-Castel, 1965).

Necrotizing pneumonia is not a condition exclusively limited to anaerobic bacteria and many aerobes can also cause this type of lesion. It is characterized by destruction and sloughing of the lung parenchyma which may be rapid and fatal (Kline & Berger, 1935) or run a chronic course.

Aspiration pneumonia can be caused by chemicals and gastric juice, solid particles or bacterial infection. Aspiration of bacteria from the oropharynx is probably a usual occurrence but the normal defence mechanisms of the respiratory tract are able to prevent infection. As shown by Smith (1930) it is possible to produce a pneumonitis followed by abscess formation by the inoculation of pyorrhoea exudate. Later Socransky and Gibbons (1965) in an experimental model showed that the presence of *B. melaninogenicus* was essential to mixtures of other naturally occurring bacteria of oral and intestinal origin to restore infectivity to the mixture. Bartlett et al. (1974a), reported that patients with aspiration pneumonia due to bacterial infection showed two features, a condition disposing to aspiration, that is altered conscious-ness, associated with general anaesthesia, drug abuse, alcoholism or central nervous system disease and radiological evidence of infection in a dependent pulmonary segment. The signs of pneumonitis preceded the development of an abscess. Anaerobes were isolated from 87 per cent of patients, this finding being confirmed in a smaller study by Swenson and Lorder (1974). Although the predominant species seen in anaerobic pneumonia are peptostreptococcus, *B. melaninogenicus, Fusobacterium nucleatum* and *B. fragilis* were present in 20 per cent of patients (Gorbach & Bartlett, 1974). Anaerobic bronchitis, although reported by Castellani (1909), probably does not exist as a specific entity and the isolation of anaerobic bacteria is a reflection of the under-lying parenchymal infection.

The bacteriology of anaerobic pleuropulmonary infections is complex and many different species may be isolated from any specimen. *S. necrophorus* and *Fusobacteria* are the most common isolates accounting for 36 per cent of the total anaerobes identified in an extensive review (Bartlett & Finegold, 1972). *B. fragilis* was reported in 9 per cent of the patients and it was most commonly associated with peritoneal infection and aspiration, although also with infection of the middle ear and mastoids.

## Upper Respiratory Tract Infections

In the past the upper respiratory tract and the ear was a major source of anaerobic infection. Gunn (1956) reported that in 54 of 173 cases, the oropharynx was the portal of entry. Tonsilitis with abscess formation and jugular vein thrombosis were common syndromes. The spectrum of upper respiratory tract infection has completely changed following the introduction of antibiotics. The widespread use of penicillin prophylactically and therapeutically has altered the causative flora, in view of its activity against anaerobic bacteria of oral origin. Anaerobes are still important pathogens in chronic sinusitis (Urdal & Berdal, 1949) and using careful techniques Frederick and Braude (1974) were able to recover anaerobic bacteria in 52 per cent of patients undergoing surgical drainage of the sinuses. Lemierre (1936) showed that otitis media and mastoiditis were important sources in anaerobic bacteraemia.

Whilst fusobacteria are probably the commonest anaerobic bacteria isolated from infections related to the mouth, bacteroides species including fragilis are frequently isolated from the sinuses and ears, and *B. melaninogenicus* is an important pathogen in gingivitis and other mouth infections (MacDonald et al., 1963).

## Central Nervous System Infections

The incidence of metastatic spread to the central nervous system in patients with bacteroides infections has been reported as between 2 and 7 per cent, although Heineman and Braude (1963) found anaerobes to be responsible for 89 per cent of brain abscesses, *B. fragilis* being isolated in 5 of the 16 cases. Ballantine and Shealy (1959) found 4 (9 per cent) of 44 patients with brain abscesses infected with bacteroides. Salibi (1964) reported the case of a young girl who had three intracranial abscesses and Lee and Berg (1971) described the first case, a newborn infant, with a cephalohaematoma infected by bacteroides, the infecting organisms in previous cases usually being *E. coli.* (Burry & Hellerstein, 1966).

Anaerobic bacteria involve the nervous system usually by direct extension from ear, sinuses and mastoid cavities or by haematogenous spread from these sites and chronic infections of the lung. Bacteroides seldom are responsible for pyogenic meningitis or post-operative infections after neurosurgery (Swartz & Karchmer, 1972) although isolated reports in meningitis have occurred (Lifshitz et al., 1963) and Ballenger et al. (1943) stated that 11 cases had been described prior to 1939.

## Soft Tissue Infections and Ulcers

Anaerobic bacteria are commonly associated with skin and soft tissue infections, especially where there is trauma and devitalization, e.g., surgical wounds or ischaemic limbs. The species responsible for the infection will depend on the primary cause, post-operative intestinal wounds, decubitus ulcer and bedsores being infected with *B. fragilis* whereas bites are infected with anaerobes of oral origin. Synergistic necrotizing cellulitis associated with multiple bacterial species, including bacteroides, is a rapid progressive lesion with necrosis of the skin and connective tissues commonly occurring in the perineum or legs, and is associated with ischaemia and diabetes and is different from the progressive bacterial synergistic gangrene of Meleney (1949) where an indolent ulceration of the skin and subcutaneous tissues occurs. The isolation of anaerobic organisms from lesions such as leg ulcers and bedsores is frequently difficult due to over-growth by large numbers of aerobic bacteria which are also present. In leg ulcers there is general disagreement regarding the role played by bacteria and in many cases bacteriological examinations are not carried out, but if they are the specimen collection techniques are not usually adequate for the isolation of anaerobes.

Pearson and Smiley (1968) reported that in pilonidal sinus while there was usually a mixed bacterial growth, bacteroides were isolated in 73 per cent of patients.

## Complications of Bacteroides Infections

Bacteroides infections are associated with a number of complications that are characteristic and diagnostically useful. Local thrombophlebitis is probably a constant feature of all bacteroides infections (Reid et al., 1945), although in many mild cases it may be overlooked. The production of heparinase by bacteroides is possibly important in the development of the thrombophlebitis and this may occur even if the patient is receiving heparin. The incidence of septic thrombophlebitis has been reported as between 5 and 20 per cent (Tynes & Frommeyer, 1962; Marcoux et al., 1970; Wilson et al., 1972).

In tonsillar abscesses which led to thrombophlebitis of the internal jugular vein, even ligation was tried (Kissling, 1929; Frankel, 1929) without success. Interestingly, some workers (Sinkovics & Smith, 1970; McHenry et al., 1961) have not found thrombophlebitis but the severity of this condition will depend on the site of the primary infection. Pulmonary embolization has been described in many cases, Nobles (1973) found that 10 of 43 cases developed pulmonary emboli, Felner and Dowell (1971) 74 of 250 patients and Bodner et al. (1970) 3 of 39 patients with bacteroides bacteraemia.

Bjornson and Hill (1973) have shown in an experimental model in mice that the injection of anaerobic bacteria caused acceleration of coagulation and this effect could be reproduced by a lipopolysaccharide extract of the cell wall of the anaerobes. The hypercoagulable state and thromboembolism seen in patients with anaerobic infections can be produced experimentally although the mechanisms are not fully understood. As a result of the embolisation with septic thrombi, metastatic abscesses are a common complication in bacteroides infection, and have been reported in 10–30 per cent of cases (Marcoux et al., 1970; Wilson et al., 1972; Nobles, 1973). The abscesses may occur anywhere in the body, but the most frequent sites are the lungs, liver, brain and joints.

Pylephlebitis is a serious complication of intestinal bacteroides infection, and most commonly follows acute appendicitis.

Although the development of metastatic abscesses may be a single episode, in many cases crops of secondary infections occur which, while not usually fatal, predispose to chronic sepsis. Jaundice, first reported by Hallé (1898) in association with bacteroides bacteraemia, can be present in 10 to 40 per cent of infections (McHenry et al., 1961; Gelb & Seligman, 1970, Wilson et al., 1972).

The mortality of bacteroides infections is primarily related to the presence of bacteraemia although age, co-existing disease and the effectiveness of the prescribed chemotherapy, are also important. Before the discovery of sulphonamides the mortality of patients with bacteraemia was over 81 per cent and this fell to 64 per cent with the use of sulphonamides, 50 per cent with the introduction of penicillin and streptomycin, and 33 per cent with multiple therapeutic regimes (Gunn, 1956). In a study of 200 bacteroides infections (Leigh, 1974) the mortality without bacteraemia was only 2·7 per cent but this was increased in the presence of bacteraemia to 33 per cent and similar findings have been reported by other workers (Tynes & Frommeyer, 1962; Ellner & Wasilauskas, 1971; Mackenzie & Litton, 1974). Both Bodner et al. (1970) and Wilson et al. (1972) found that in the presence of shock the mortality rose to over 50 per cent and Kagnoff et al. (1972) reported a mortality of 71 per cent in their series but in only 35 per cent was death closely related to the infection. Chow et al. (1974) showed that where appropriate antibiotic treatment was given, the mortality fell from 62 to 16 per cent. The morbidity associated with bacteroides infections is considerable, as even with further surgical procedures and appropriate chemotherapy the lesion may only resolve slowly necessitating long periods of hospitalization or absence from work. Leigh (1974) reported an incidence of complications of nearly 30 per cent in bacteroides infections seen in a district hospital. It is essential to prescribe therapy for in many cases of formed bacteroides abscess, although surgical drainage may be the treatment of choice, the antibiotics may prevent the development of metastatic abscesses by sterilizing the embolised septic thrombi. Because of the septic thrombophlebitis surrounding bacteroides infections, it is wise to give prophylactic treatment over the period of surgery, especially where intra-abdominal infections are present.

In neonates, although Pearson and Anderson (1967) reported uniform mortality, later workers (Chow et al., 1974) found a mortality of only 4 per cent, the only death occurring in a group of 9 untreated infants, and reviewing the previous literature found a mortality of 26 per cent. There is little doubt that severe infections with anaerobic organisms do occur in neonates and are associated with serious mortality, but transient bacteraemia can occur which follows a relatively benign course and may be primarily related to infected amniotic fluid in the mother.

## Antibiotic Sensitivities

The bacteroides group of organisms are unusual and differ from the genera of the Enterobacteriaciae in their susceptibility to antibiotics. Although there is sensitivity to many of the oral antibiotics such as tetracycline and erythromycin, the substances normally used to treat severe infections such as the penicillins and aminoglycosides are without therapeutic activity. Indeed, many chemotherapeutic substances introduced into clinical pharmacology because of their Gram-positive spectrum of activity have later been found to have a considerable effect on bacteroides.

The determination of the antibiotic susceptibility of bacteroides is not without problems. Ingham et al. (1970) have shown that the effect of 10 per cent $CO_2$ in the culture environment alters the sensitivity results for certain antibiotics and with erythromycin and lincomycin the minimal inhibitory concentrations may be up to 32 times higher. However, the presence of $CO_2$ stimulates the growth of anaerobes and

**Table 5.2.** *Antibiotic susceptibility of Anaerobic Bacteria.*

| | Sphaerophorus necrophorus | Fuso-bacterium | Bacteroides fragilis | Bacteroides melaninogenicus |
|---|---|---|---|---|
| Penicillin G (2 units) | S | S | R | S |
| Erythromycin (60 µg) | V | R | S | S |
| Kanamycin (1000 µg) | R | S | R | R |
| Vancomycin (5 µg) | R | R | R | V |
| Colistin (10 µg) | S | S | R | V |
| Rifampicin (15 µg) | S | R | S | S |

may allow an in vitro reaction between antibiotic and bacteria to occur at a rate similar to the in vivo situation.

The methods for determining the antibiotic sensitivity of *B. fragilis* are similar to those for any organism. An indication of susceptibility can be determined using the impregnated disc technique, and several agars have been advocated which allow the growth of fastidious anaerobes (Sutter et al., 1972; Barry & Fay, 1974). Although the method in use in most clinical laboratories, disc sensitivity testing, is open to many errors and can only provide a rough guide, in most cases this is sufficient for therapeutic indications.

A more accurate method is to determine the minimal inhibitory concentration by the tube or plate dilution methods. The tube method is the least convenient as handling during anaerobic incubation is difficult. The best method is to incorporate dilutions of the antibiotic into agar which will support the growth of *B. fragilis* but will not inter-fere with the action of the antibiotic. Prolonged incubation for up to 72 h is needed and the bactericidal effect can be detected by 'imprint' methods onto agar without antibiotic. The bacterial inoculum is im-portant with certain antibiotics and a concentration between $10^5$ and $10^6$ organisms per ml is usually used.

## The Penicillins

Whilst penicillin is the antibiotic of choice against the bacteroides species found in the mouth, *B. fragilis* usually shows a level of resistance which is beyond the clinical therapeutic range. Of the newer penicillins, ampicillin and carbenicillin shows the greatest activity (Blazevic & Matsen, 1974), the latter only because of the higher blood concentrations that can be obtained on systemic treatment. In clinical practice a proportion of strains of *B. fragilis* are susceptible to ampicillin on disc sensitivity testing and using systemic therapy about 50 per cent could be expected to respond, providing blood concentrations in the range 12·5–25·0 mg/l are attained. In routine practice however, the penicillins are not indicated in the treatment of *B. fragilis* infections.

## The Cephalosporins

Like the penicillins, the cephalosporins show a low level of activity against bacteroides on a weight for weight basis, and to treat at least 50 per cent of the strains requires blood concentrations considerably in excess of 25 mg per l (Kislak, 1972). Until the recent introduction of the new cephalosporins, cephaloridine showed the greatest activity, but cephazolin is effective for many strains of *B. fragilis* by virtue of the higher blood levels attained on systemic therapy, although on a weight for weight basis it is no more active than cephaloridine, or cephalothin. The oral cephalosporins (Leigh & Simmons, 1975a) at present in clinical use are not very effective, although cephradine is more active than cephalexin.

## Erythromycin

The initial reports (Garrod, 1955; Finegold & Hewitt, 1956) suggested that erythromycin was active against a high proportion of strains of bacteroides, and Ingham et al., (1968) described the successful treatment of a case of severe infection. Although the median MIC reported by Kislak (1972) was only 1·56 mg per l, the blood level attained on oral administration is only between 2 and 3 mg per l, and Leigh (1974) found less than 30 per cent of strains susceptible to this level. Systemic therapy with this compound has difficulties which limits its usefulness considerably.

## Chloramphenicol

Many workers have reported that chloramphenicol shows considerable activity against strains of *B. fragilis* (Finegold & Hewitt, 1956; Saksena et al., 1968; Ellner & Wasilauskas, 1971), only 10 per cent being resistant to 12·5 mg per l (Martin et al., 1972; Bodner et al., 1972). However there is some inconsistency in the antibacterial effect, although this may not be important clinically. The main disadvantage of chloramphenicol is its potential toxicity to the bone marrow in adults and the considerable risks of the 'gray' syndrome in premature babies.

Although it has been used with considerable success in the past, there have been reports of significant failures (Haldane & van Rooyen, 1972; Bartlett et al., 1972; Thadepalli et al., 1973b) and chloramphenicol is now rarely used as the antibiotic of choice.

## The Tetracyclines

For many years the tetracyclines were a first line therapy for bacteroides infections. The distribution of the minimal inhibitory concentrations shows approximately 40 per cent of strains with a high level of sensitivity (Kislak, 1972; Leigh, 1974) and a second group of strains resistant to 10 mg per l. It appears that the number of strains in the resistant portion of the curve is increasing and includes up to 65 per cent of strains (Sutter et al., 1972). Ingham et al., (1968) reported the failure of tetracycline to exert a consistent bactericidal effect. The newer analogues of tetracycline, such as doxycycline and minocycline, show a greater activity against bacteroides (Leigh & Simmons, 1975b) over 90 per cent of strains being susceptible to normal blood levels of minocycline, and 75 per cent for doxycycline, although the bimodal distribution of sensitivity is still present. The clinical use of these compounds is limited at the present time.

## Lincomycin

Lincomycin introduced for the treatment of Gram-positive infections is active against most strains of *B. fragilis* (Lewis et al., 1963) although systemic therapy is necessary to attain sufficiently high blood concentrations. The median MIC has been reported at 3·1 mg per l Kislak, 1972) but only levels of up to 7 mg per l are attained on oral administration. The reports of the susceptibility of bacteroides to lincomycin has varied considerably from over 90 per cent (Bodner et al., 1972) to less than 50 per cent (Chow & Guze, 1971). Lincomycin has been used with good results in the treatment of bacteroides infections and has revolutionised the therapeutic regime, but has now been superseded by clindamycin in view of its considerably greater activity against anaerobic organisms.

## Clindamycin

Clindamycin is a chlorine derivative of lincomycin, and although introduced primarily for Gram-positive infections has excellent activity against all anaerobes. Many studies have demonstrated clearly that over 90 per cent of Bacteroides are susceptible to concentrations readily attainable in the blood (Chow & Guze, 1971; Bodner et al.. 1972, Bartlett et al., 1972), and in others no untreatable strains have been found (Nobles, 1973; Nastro & Finegold, 1972; Leigh, 1974). Although not bactericidal in all cases, clindamycin is not influenced by either the presence of an alkaline pH or by the use of carbon dioxide in the cultural environment. Clindamycin is, at the moment, the antibiotic of choice in the treatment of *B. fragilis* infections.

## Rifampicin

Bacteroides show a marked susceptibility to rifampicin and the majority of strains have been shown to be sensitive to low concentrations such as 0·3 mg per l (Ingham et al., 1970; Werner, 1972; Finegold et al., 1970b). However, there are problems of the 'skip tube' phenomenon where rapid one step resistance occurs, Nastro and Finegold (1972) noted this in 4 of 19 strains tested, and an inconsistent bactericidal effect was found in 50 per cent of strains tested by Ingham et al., (1970). It is likely that the use of rifampicin alone in bacteroides infections would not be a sound or successful therapy, and it should probably be reserved for the treatment of tuberculosis. In England this compound has been withdrawn from the market except as a preparation for tuberculosis therapy.

## Metronidazole

Metronidazole has been used extensively in the treatment of *Trichomonas vaginalis* infections (Durel et al., 1960), but for some years its activity against anaerobic bacteria was not known, until Shinn (1962) observed that it was active against fusobacteria. Prince et al. (1969) demonstrated in vitro inhibition of strains of clostridium, fusobacterium, and veillonella, and Fuzi and Csukás (1970) activity against strains of Bacteroides species, and *Fusobacterium fusiforme*. Clinical effectiveness of metronidazole has been shown in 'Vincents angina' or acute ulcerative gingivitis (Shinn et al., 1965). The minimal inhibitory concentrations are usually between 0·16 and 2·5 mg per l (Whelan & Hale, 1973) and a consistent bactericidal effect is achieved (Nastro & Finegold, 1972) and may be effective in treating severe infections such as endocarditis, providing oral therapy is possible. Metronidazole has the advantage that its activity against obligatory anaerobic organisms is specific and there is no effect on aerobic bacteria. High blood levels are achieved, a single 2 g dose producing a serum level of 46 mg per l, which falls slowly so that at 48 h it is still 3·0 mg per l. Even on a low dose of 200 mg the level attained is 5 mg per l (Kane et al., 1961) and Leigh (1974) has shown this to be effective against 96 per cent of strains isolated from clinical material an intravenous preparation is now also being evaluated.

## The Aminoglycosides

The aminoglycoside group of antibiotics (Gentamicin, Kanamycin, Neomycin), have no role in the therapy of bacteroides infections. The minimal inhibitory concentrations in many cases exceed 3000 mg per l even with gentamicin (Finegold & Sutter, 1971) and this level probably cannot even be achieved by topical application or in the gastrointestinal tract by oral therapy. The use and efficacy of these compounds as a bowel preparation for sterilization is therefore greatly questionable, as over 99 per cent of the faecal flora will be resistant. Indeed, the amino-glycosides form part of 'clinical laboratorys' isolation procedure for anaerobic bacteria, as usually neomycin or kanamycin are routinely incorporated into diagnostic agar plates to inhibit the growth of other bacteria. In addition, as strains of anaerobic bacilli, other than *B. fragilis*, show a greater range of susceptibility (Finegold et al., 1967), this activity can be used by the routine laboratory in the identification of these organisms.

## Spectinomycin

This antibiotic is an aminocyclitol not an aminoglycoside, but is very unusual in that it shows activity against *B. fragilis*. Peak serum levels attained after normal dosage are very high, and exceed the resistance of most strains (Phillips & Warren, 1975; Leigh, 1975). Clinical trials have not yet been carried out to prove the in vitro findings.

## Fusidic Acid

Although introduced as an antibiotic specifically indicated for infections due to Gram-positive cocci, particularly *Staph. aureus*, fusidic acid has considerable activity against strains of *B. fragilis*. Seventy per cent of strains isolated from clinical infections being sensitive to concentrations attainable in the blood on standard dosage. As yet no clinical trials have been carried out in bacteroides infections, but it may be found that this is a useful antibiotic for the local treatment of superficial bacteroides infections or abscesses, especially as a 'single shot' therapy (Caviject) is available for use in abscess cavities.

## Nitrofurantoin

Nitrofurantoin is a chemotherapeutic agent which is used mainly for urinary tract infections, but it shows considerable activity against *B. fragilis* (Tynes & Frommeyer, 1962; Schoutens et al., 1973). It might therefore be useful in the treatment of bacteroides infections of the urinary tract, but these are rare and the poor blood concentrations will exclude infections in the tract that are not accessible to the high urinary concentration.

## Treatment

The treatment of infections due to bacteroides depends on the type of infection that is present. A majority of infections are abscesses or closed space infections such as empyema or arthritis and the primary therapy in these cases is surgical drainage and removal of necrotic and devitalized tissue. Although a surgical approach is essential to the correct management, unlike many other bacterial abscesses, the use of chemotherapy is also of considerable importance, as it may prevent the common complication of metastatic abscess formation. The use of appropriate antibiotic treatment has been shown to have a striking effect on the incidence of mortality and morbidity. Bacteraemia due to bacteroides is an automatic indication for antibiotic treatment and it is important that the substances chosen have a great activity against bacteroides or else the incidence of mortality and other complications is high. In all bacteroides infections the importance of chemotherapy can be judged from the rate of complications seen with inappropriate treatment.

Penicillin, the antibiotic of choice in infections due to most anaerobic bacteria, is not effective in *B. fragilis* infections, although on rare occasions a good clinical response has been achieved on high dosage, and in these cases the strains must have been relatively sensitive. In general, the other penicillins show little activity but where high blood and tissue concentrations can be achieved by continuous intravenous administration, some clinical success can be attained. Chow and his co-workers (1974) have suggested that penicillin and ampicillin are more effective in neonatal bacteraemia. The cephalosporins have been found to be successful in some cases particularly cephalothin, (Marcoux et al., 1970) and it is possible that the newer systemic preparations such as cefazolin may be effective (Leigh & Simmons, 1975a).

In the past, sulphonamides have been used with good results in animals (Prevot, 1940) and human infections (Tynes & Frommeyer,

1962; Felner & Dowell, 1971) either alone or in combination with a tetracycline. Many strains of *B. fragilis* are still susceptible to sulphonamides, but it is unlikely that these substances would work successfully in tissue infections. However, they might be more effective than the aminoglycosides in suppressing the faecal flora in pre-operative bowel preparations. Cotrimoxazole, a combination of trimethoprim and sulphamethoxazole, shows considerable activity in vitro against strains of *B. fragilis* (Okubadejo et al., 1974), but there are no reported clinical studies of its use. However at High Wycombe and Amersham hospitals it has been found to be effective in the treatment of postpartum pyrexia due to intrauterine infection, where no significant pathogens have been isolated from vaginal swabs. In these cases the incidence of anaerobic infections is high.

Following the introduction of the tetracyclines and chloramphenicol, there have been two main schools of therapy advocating these substances. In view of its lesser toxicity most workers have suggested tetracycline and have achieved significant success (Gillespie & Guy, 1956; Vinke & Borghans 1963; Marcoux et al., 1970; Sinkovics & Smith, 1970; Gelb & Seligman, 1970; Ledger et al., 1971; Kagnoff et al., 1972). Hermans and Washington (1970) used tetracycline in a triple antibiotic regime in the treatment of polymicrobial bacteraemia and Schatten (1956) intraperitoneally in peritonitis. However up to 50 per cent of strains of *B. fragilis* now show significant resistance to tetracycline and its efficacy is in question. The increased activity of the new compounds such as minocycline and doxycycline has not been proved by clinical trial.

Chloramphenicol has been advocated by many workers (Lifshitz et al., 1963; Rein & Cosman, 1971; Masri & Grieco, 1972; Baird, 1973) and shows excellent results. The toxicity to the bone marrow, however, and the reported poor bactericidal effect in vitro with some strains has led most workers to use this as a second line antibiotic.

The treatment of choice in severe bacteroides infections is at present either clindamycin or lincomycin. In vitro studies show that especially clindamycin has exceptional activity whether used orally or systemically, and lincomycin can achieve similar results on systemic usage. The systemic preparation of clindamycin is only now available in the UK, but reports from America suggest that nearly all patients with anaerobic infections including *B. fragilis* show a rapid response to therapy (Bartlett, et al., 1972; Haldane & van Rooyen, 1972; Shumer et al., 1973; Fass et al., 1973; Chow et al., 1974). Gorbach and Thadepalli (1974) found a satisfactory response from systemic clindamycin alone for mixed aerobic/anaerobic infections, and suggested that this indicated the major significance of anaerobes. In anaerobic pneumo-pulmonary infections Sen et al. (1974) found that 26 of 30 patients showed a favourable response, and using lincomycin Ziment et al. (1969) showed efficacy in joint infections and Geddes et al. (1967); Tracy et al. (1972) in abdominal infections. In genital tract infections, clindamycin has resulted in clinical and bacteriological cure (Thadepalli et al., 1973a; Ledger et al., 1974).

The range of oral treatments for *B. fragilis* infections has recently been enlarged by reports of the use of metronidazole. Willis and his co-workers (1974) have shown this compound to be effective both in the therapy of established infections and in the reduction of carriage of bacteroides in the genital tract. The consistent bactericidal action has led Nastro & Finegold (1973) to suggest the use of metronidazole in bacterial endocarditis where previously, with other chemotherapeutic agents, only systemic therapy provided reliable blood concentrations. The use of intravenous metronidazole has recently been reported by Selkon et al., (1975) and the availability of this preparation will be valu-able to the treatment of severe anaerobic infections although its limited range of bacterial activity will necessitate combined therapy. However, in laboratory studies no adverse interactions with 14 commonly used antibiotics have been found (Salem et al., 1975).

Although used rarely now, erythromycin has been prescribed for bacteroides infections with success (Ingham et al., 1968; Baird, 1973) but difficulties with systemic administration, toxicity and recent reports of increased resistance limit its use. Rifampicin would be very effective in the treatment of biliary infections, but it has unfortunately been withdrawn from the market. Fusidic acid while showing considerable activity in vitro has not been used in controlled clinical trials, and nitrofurantoin, although effective against many strains of *B. fragilis* (Schoutens et al., 1973) and possibly useful in infections of the urinary tract, is unlikely to be considered in view of the poor blood levels and the possibility of bacteraemia.

## Conclusion

Although there have been unaccountable lapses of interest in the importance and frequency of bacteroides infections since the first descriptions by Veillon and Zuber (1898) and Hallé (1898), there is now, in the late sixties and early seventies, considerable interest and some workers have described a new 'bandwagon' approach to these infec-tions. Whilst previous upsurges of interests, for example concerning staphylococci and asymptomatic bacteriuria, have either been an exaggerated problem, most hospitals do not see epidemics of staphylococcal infection, or have remained unsolved, the characteristics associated with susceptibility to urinary infection are still little known, the situation appears different with anaerobic infections. *B. fragilis* is present in the faecal flora of every person and presents a continuing hazard to all patients who develop intestinal diseases, conditions of the genital tract, and pregnancy. There is a new type of difficulty associated with the importance of bacteria that are rarely isolated by the routine laboratory, and require a conscious effort in terms of specimen collection, transport and laboratory culture by hospital staff; whilst it is unacceptable that 'sterile pus' occurs in the absence of concurrent chemotherapy it is also difficult to ignore large numbers of easily cultivated aerobic bacteria and look for anaerobic organisms. It is thought by some workers that the whole reality of anaerobic isolation is a hoax, as these organisms are frequently found in mixed culture and the evidence for pathogenicity is poor. However, there are several significant findings which are now being universally reported. Careful examination techniques of sterile pus and fluid reveals the presence of anaerobes and in many studies these samples may account for 50 per cent of specimens. Where contamination due to perforation of an organ such as the colon or appendix occurs, although primary culture may reveal a mixed aerobic/anaerobic flora, the development of a metastatic abscess or post-operative wound infection usually shows a high incidence of pure anaerobes. In mixed cultures the relative numbers of bacteria show a considerably greater count of anaerobes, although frequently on laboratory culture the reverse would appear true. Recent studies into bacteraemia, which is accepted as strong evidence of pathogenicity, have shown that *B. fragilis* is the second most commonly isolated organism after *E. coli.* The suppression of anaerobes by specific chemotherapeutic agents such as clindamycin or metronidazole is associated with a reduction in incidence of infective complications of surgery or injury, although other aerobic organisms may still be present. Perhaps the most significant evidence for the pathogenicity of bacteroides is through the accidental treatment of mixed infections by chemotherapeutic agents that only exert an effect on anaerobes. Many workers have documented cases where even though anaerobes are combined with other bacterial species such as *E. coli, Pseudomonas aeruginosa, Klebsiella* and *Enterobacter,* a cure was effected by the use of clindamycin as the single agent. At the moment the numbers are too small to make recommendations, and most workers would still advocate the use of antibiotics active against all bacteria isolated. Aerobic bacteria, such as *E. coli* and streptococci, are responsible for a significant number of infections resulting from injury or surgery to the intestinal tract, and it is a difficult decision to ignore these organisms on the assumption that the anaerobes present have failed to grow. It may be many years before the relative importance of the anaerobic and aerobic components of a severe mixed infection can be judged.

Before any universal decision can be taken, it is necessary for all laboratories to have adequate facilities for anaerobic culture and for all clinicians to be aware of the specific characteristics of anaerobic infection so that the vital transport precautions may be undertaken at an early stage in the diagnosis. The characteristics that are helpful in the diagnosis are the site of infection, the nature of the pus or exudate, particularly the odour, and the response of previous antibiotic therapy.

The eventual control and prevention of *B. fragilis* infections as with all other infections, depends on the recognized frequency as a complication of particular conditions and diseases and the use of specific prophylactic therapy. Both the clinician and the laboratory still have a lot to learn but the careful studies over the last few years have provided a sound base on which to develop new techniques of diagnosis and treatment.

## References

Alling, B., Brandberg, A., Seeberg, S. & Svanborg, A. (1973) Aerobic and anaerobic microbial flora in the urinary tract of geriatric patients during long term care. *J. infect. Dis.,* **127**, 34.

Altemeier, W. A. (1938) The cause of the putrid odour of perforated appendicitis with peritonitis. *Ann. Surg.,* **107**, 634.

Altemeier, W. A., Culbertson, W. R. & Fullen, W. D. (1973) Intra-abdominal abscesses. *Am. J. Surg.* 125, 70.

Ament, M. E. & Gaal, S. A. (1967) Bacteroides arthritis. *Am. J. Dis. Child.,* **114**, 427.

Anaerobic Laboratory, Virginia Polytechnic Institute (1970) *Outline of clinical methods in anaerobic bacteriology.* Blacksberg, Virginia.

Anderson, J. D. & Sykes, R. B. (1973) Characterisation of a β lactamase obtained from a strain of Bacteroides fragilis resistant to β lactam antibiotics. *J. med. Microbiol.,* **6**, 201.

Andrews, E. & Henry, L. D. (1935) Bacteriology of normal and diseased gall bladder. *Archs. intern. Med.,* **56**, 1171.

Attebery, H. R. & Finegold, S. M. (1969) Combined screwcap and rubber stopper closure for Hungate tubes (pre-reduced, anaerobically sterilised roll tubes and liquid media). *Appl. Microbiol.,* **18**, 558.

Baird, R. M. (1973) Post-operative infections from Bacteroides. *Am. Surg.,* **39**, 459.

Ballantine, H. T. & Shealy, C. N. (1959) Role of radical surgery in the treatment of abscess of the brain. *Surg. Gynae. Obst.,* **109**, 370.

Ballenger, J., Schall, L. A. & Smith, W. E. (1943) Bacteroides meningitis. *Ann. Otol.,* **52**, 895.

Barnes, E. M., Impey, C. S. & Goldberg, H. S. (1966) Methods for the characterisation of the Bacteroidaciae. In: *Identification methods for Microbiologists,* Part A, p. 51, Eds, E. M. Gibbs & F. A. Skinner, London: Academic Press.

Barnes, E. M. & Goldberg, H. S. (1968) The relationships of bacteria within the family Bacteroidaciae as shown by numerical taxonomy. *J. gen. Microbiol.,* **51**, 313.

Barry, A. L. & Fay, G. D. (1974) Evaluation of four disc diffusion methods for antimicrobic susceptibility tests with anaerobic gram negative bacilli. *Am. J. clin. Path.,* **61**, 592.

Bartlett, J. G. & Finegold, S. M. (1972) Anaerobic pleuropulmonary infections. *Medicine,* **51**, 413.

Bartlett, J. G., Sutter, V. L. & Finegold, S. M. (1972) Treatment of anaerobic infections with lincomycin and clindamycin. *New Eng. J. Med.,* **287**, 1006.

Bartlett, J. G., Gorbach, S. L. & Finegold, S. M. (1974a) The bacteriology of aspiration pneumonia. *Am. J. Med.,* **56**, 202.

Bartlett, J. G., Gorbach, S. L., Tally, F. B. & Finegold, S. M. (1974b) Bacteriology and treatment of primary lung abscess. *Am. Rev. resp. Dis.,* **109**, 510.

Beerens, H., Schaffner, Y., Guillaume, J. & Castel, M. M. (1963) Les bacilles anaérobies non sporulés à Gram négatif favorisés par la bile leur appartenance au genre Eggerthella (nov gen). *Ann. Inst. Past.,* **14**, 5.

Beerens, H. & Tahon-Castel, M. (1965) Infections humaine à bactéries anaérobies non toxigènes. Brussels, Presse Acad Europ. 1965.

Beerens, H. (1970) Report of the International Committee for Gram nomenclature of bacteria. Taxonomic Sub-Committee for Gram negative anaerobic rods. *Int. J. Syst. Bacteriol.,* **20**, 297.

Beigelman, P. M. & Rantz, L. A. (1949) Clinical significance of Bacteroides. *Arch. intern. Med.* **84**, 605.

Bergey, D. H., Harrison, F. C., Breed, R. S., Hammer, B. W. & Muntoon, F. M. (1923) Bergey's manual of Determinative Bacteriology. Baltimore, Williams & Wilkins.

Bjornson, H. S. & Hill, E. O. (1973) Bacteroidaciae in thromboembolic disease: effects of cell wall components on blood coagulation in vivo and vitro. *Infect. Immun.* **8**, 911.

Blazevic, D. J. & Matsen, J. M. (1974) Susceptibility of anaerobic bacteria to carbenicillin. *Anti-microb. Ag. Chemother.,* **5**, 462.

Bodner, S. J., Koenig, M. G. & Goodman, J. S. (1970) Bacteraemic bacteroides infections. *Ann. int. Med.,* **73**, 537.

Bodner, S. J., Koenig, M. G., Treanor, L. L. & Goodman, J. S. (1972) Antibiotic susceptibility testing of Bacteroides. *Anti-microb. Ag. Chemother.,* **2**, 57.

Bornstein, D. L., Weinberg, A. N., Swartz, M. N. & Kunz, L. J. (1964) Anaerobic infections—review of current experience. *Medicine,* **43**, 207.

Breed, R. S., Murray, E. G. D. & Smith, N. R. Eds (1957) Bergey's manual of Determinative Bacteriology. 7th ed. London: Baillière, Tindall & Cox.

Brewer, J. H. (1939) A modification of the Brown anaerobe jar. *J. Lab. clin. Med.,* **24**, 1190.

Brewer, J. H., Heer, A. A. & McLaughlin, C. B. (1955) The use of sodium borohydride for producing hydrogen in an anaerobic jar. *Appl. Microbiol.,* 3, 136.

Brown, A. E., Williams, H. L. & Herrell, W. G. (1941) Bacteroides septicaemia. *J. Am. med. Assoc.,* 116, 402.

Brown, J. H. (1921) An improved anaerobe jar. *J. exp. Med.,* 33, 677.

Brown, J. H. (1922) Modification of an improved anaerobe jar. *J. exp. Med.,* 35, 467.

Brown, T. K. (1930) The incidence of puerperal infection due to anaerobic streptococci. *Am. J. Obstet. Gynec.,* 20, 300.

Brown, W. R. & Lee, E. (1974) Radioimmunological measurements of bacterial antibodies. ii. Human serum antibodies reactive with Bacteroides fragilis and Enterococcus in gastrointestinal and immunological disorders. *Gastroenterology.* 66, 1145.

Brumfitt, W. & Leigh, D. A. (1969) The incidence and bacteriology of bacteraemia—a study at two hospitals. *Proc. R. Soc. Med.,* 62, 1239.

Burdon, K. L. (1928) Bacterium melaninogenicum from normal and pathological tissues. *J. infect. Dis..* 42, 161.

Burry, V. F. & Hellerstein, S. (1966) Septicaemia and subperiosteal cephalohaematoma. *J. Pediat.,* 69, 1133.

Butler, T. J. and McCarthy, C. F. (1969) Pyogenic liver abscess. *Gut,* 10, 389.

Buttiaux, R., Beerens, H. & Tacquet, A. (1962) *Manual de techniques bactériologiques.* 1st ed., p. 505, Paris: Flammarion.

Buttiaux, R., Beerens, H. & Tacquet, A. (1966) *Manual de techniques bactériologiques.* 2nd ed., p. 572, Paris: Flammarion.

Buttiaux, R., Beerens, H. & Tacquet, A. (1969) *Manual de techniques bactériologiques.* 3rd ed., p. 707, Paris: Flammarion.

Carter, B., Jones, C. P., Ross, R. A. & Thomas, W. L. (1951) A bacteriologic and clinical study of pyometria. *Am. J. Obstet. Gynec..* 62, 793.

Carter, B., Jones, C. P., Alter, R. A., Creadick, R. N. & Thomas, W. L. (1953) Bacteroides infections in obstetrics and gynaecology. *Obstet. Gynec. N.Y.,* 1, 491.

Castellani, A. (1909) Bronchial spirochaetosis. *Br. med. J.,* 2, 782.

Castellani, A. & Chalmers, A. J. (1919) Manual of Tropical Medicine. 3rd ed. New York: Balliere, Tindall & Cox.

Castleman, B. & McNeely, B. U. (1970) Case records of Massachusetts General Hospital. Case 27-1970. *New. Eng. J. Med.,* 282, 1477.

Cates, J. & Christie, R. V. (1951) Subacute bacterial endocarditis: review of 442 patients treated in 14 centres appointed by Penicillin Trials Committee of the Medical Research Council. *Q. J. Med.,* 20, 193.

Cato, E. B., Cummins, C. S., Holdeman, I. V., Johnson, J. I., Moore, W. E. C., Smibert, R. M. & Smith, L. D. S. (1970) Outline of clinical methods in anaerobic bacteriology. 2nd ed., p. 107, Blacksberg, Virginia: Virginia Polytechnic Institute.

Chandler, F. A. & Breaks, V. M. (1941) Osteomyelitis of the femoral neck and head. *J. Am. med. Assoc.,* 116, 2390.

Chow, A. W. & Guze, L. B. (1971) More on antibiotic susceptibility of Bacteroides. *Ann. int. Med.,* 75, 810.

Chow, A. W., Leake, R. D., Yamauchi, T., Anthony, B. E. & Guze, L. B. (1974) The significance of anaerobes in neonatal bacteraemia: analysis of 23 cases and review of literature. *Pediatrics,* **54**, 736.

Chow, A. W., Montgomerie, J. Z. & Guze, L. B. (1974) Parenteral clindamycin phosphate therapy for severe anaerobic infections, a clinical and laboratory evaluation. *Archs. intern. Med.,* **134**, 78.

Chow, A. W. & Guze, L. B. (1975) Bacteroidaciae bacteraemia, clinical experience with 112 patients. *Medicine, Baltimore.* **53**, 93.

Clark, C. E. & Wiersma, A. F. (1952) Bacteroides infections of the female genital tract. *Am. J. Obstet. Gynec.,* **63**, 371.

Cohen, J. (1932) The bacteriology of abscess of the lung and methods for study. *Archs. Surg.,* **24**, 171.

Collee, J. G., Watt, B., Fowler, E. B. & Brown, R. (1972) An evaluation of the Gaspak system in the culture of anaerobic bacteria. *J. appl. Bact.,* **35**, 71.

Crowley, N. (1970) Some bacteraemias encountered in hospital practice. *J. clin. Path.,* **23**, 166.

de la Cruz, E. & Cuadra, C. (1969) Antigenic characteristics of five species of human Bacteroides. *J. Bact.,* **100**, 1116.

Dack, G. M. (1940) Non-sporing anaerobic bacteria of medical importance. *Bact. Rev.,* **4**, 227.

Davis, C. E., Hunter, W. J., Ryan, J. L. & Braude, A. I. (1973) Simple method for culturing anaerobes. *Appl. Microbiol.,* **25**, 216.

Dixon, C. F. & Deuterman, J. L. (1937) Post-operative bacteroides infections, report of 6 cases. *J. Am. med. Assoc.,* **108**, 181.

Dowell, V. R., Hill, E. O. & Altemeier, W. A. (1964) Use of phenethyl alcohol in media for the isolation of anaerobic bacteria. *J. Bact.,* **88**, 1811.

Drasar, B. S., Hill, M. J. & Shiner, M. (1966) The deconjugation of bile by human intestinal bacteria. *Lancet,* **1**, 1237.

Drasar, B. S., Hughes, W. H., Williams, R. E. O. & Shiner, M. (1966) Bacterial flora of the normal intestine. *Proc. R. Soc. Med.,* **59**, 1243.

Drasar, B. S. (1967) Cultivation of anaerobic intestinal bacteria. *J. Path. Bact.* **94**, 417.

Durel, P., Roiron, V., Siboulet, A. & Burel, L. J. (1960) Systemic treatment of human trichomoniasis with a derivative of nitroimidazole, 8823. RP. *Br. J. vener. Dis.,* **36**, 21.

Edlund, V. A., Mollstedt, B. O. & Ouchterlony, O. (1958) Bacteriological investigation of the biliary system and liver in biliary tract disease correlated to clinical data and microstructure of the gall bladder and liver. *Acta. clin. Scand.,* **116**, 461.

Eggerth, A. H., & Gagnon, B. H. (1933) The Bacteroides of human faeces. *J. Bact.,* **25**, 389.

Ellner, P. D. & Wasilauskas, B. L. (1971) Bacteroides septicaemia in older patients. *J. Am. Geriat. Soc.,* **19**, 296.

Evans, J. M., Carlquist, P. R. & Brewer, J. H. (1948) A modification of the Brewer anaerobic jar. *Am. J. clin. Path.,* **18**, 745.

Fass, R. J., Scholand, J. F., Hodges, G. R. & Saslaw, S. (1973) Clindamycin in the treatment of serious anaerobic infections. *Ann. intern. Med.,* **78**, 853.

Felner, J. H. & Dowell, V. R. (1971) Bacteroides bacteraemia. *Am. J. Med.,* **50**, 787.

Finegold, S. M. & Hewitt, W. L. (1956) Antibiotic sensitivity pattern of Bacteroides species. *Antibiotics. A.,* **3**, 794.

Finegold, S. M., Miller, L. G., Merrill, S. L. & Posnick, D. (1965) Significance of anaerobic and capnophilic bacteria isolated from the urinary tract. *In: Progress in Pyelonephritis,* E. H. Kass, Ed. Philadelphia: F. A. Davis & Co., p. 159.

Finegold, S. M., Harada, N. E. & Miller, L. G. (1967) Antibiotic susceptibility patterns as aids in classification and characterisation of gram negative anaerobic bacilli. *J. Bact.,* **94**, 1443.

Finegold, S. M. & Miller, L. G. (1968) Susceptibility to antibiotics as an aid in classification of gram negative anaerobic bacilli. *In: The anaerobic bacteria,* p. 139, V. Fredette, Ed. P.Q. Canada: Montreal University.

Finegold, S. M., Sutter, V. L., Cato, E. P. & Holdeman, L. V. (1970a) Anaerobic bacteria. *In: Rapid diagnostic methods in medical microbiology.* C. D. Graber, Ed., p. 48. Baltimore: Williams and Wilkins.

Finegold, S. M., Sutter, V. L. & Sugihara, P. T. (1970b) Susceptibility of anaerobic bacteria to rifampicin. *Bacteriol. Proc.,* **73**, 81.

Finegold, S. M. & Sutter, V. L. (1971) Susceptibility of gram negative bacilli to gentamicin and the aminoglycosides. *J. infect. Dis.,* **124**, Suppl. 56.

Finegold, S. M. & Rosenblatt, J. E. (1973) Practical aspects of anaerobic sepsis. *Medicine,* **52**, 311.

Frankel, E. (1929) quoted by Gunn, A. A.

Frederick, J. & Braude, A. I. (1974) Anaerobic infections of the paranasal sinuses. *New. Eng. J. Med.,* **290**, 135.

Fredericq, P. & Levine, M. (1947) Antibiotic inter-relationship among the enteric group of bacteria. *J. Bacteriol.,* **54**, 785.

Freter, R. (1955) The fatal enteric cholera infection in the guinea pig achieved by inhibition of normal flora. *J. infect. Dis.,* **97**, 57.

Friedberg, C. K., Goldman, H. M. & Field, L. E. (1951) Study of bacterial endocarditis: comparisons in 95 cases. *Archs. intern. Med.,* **107**, 6.

Fukunaga, F. H. (1973) Gall bladder bacteriology, histology and gall stones, study of unselected cholecystectomy specimens in Honolulu. *Arch. Surg.,* **106**, 169.

Futch, C., Zikria, B. A. & Neu, H. C. (1973) Bacteroides liver abscess. *Surgery, St. Louis,* **73**, 59.

Fuzi, M. & Csukás, Z. (1970) Das antibakterielle Wirkungsspektrum de Metronidazole. *Zbl. Bakt. I. Ref.,* **213**, 258.

Garrod, L. P. (1955) Sensitivity of four species of Bacteroides to antibiotics. *Br. med. J.,* **2**, 1529.

Geddes, A. M., Munro, J. F., Murdoch, J. McC., Begg, K. J. & Burns, B. A. (1967) *In:* Proceedings of the 5th International Congress of Chemotherapy. K. H. Spitzy & H. Haschek, Eds. Vol. 1, p. 361. Vienna: Wiener Medizinische Akademie.

Gelb, A. F. & Seligman, S. J. (1970) Bacteroidaciae bacteraemia; effect of age and focus of infection on clinical course. *J. Am. med. Assoc.,* **212**, 1038.

Gesner, B. M. & Jenkin, C. R. (1961) Production of heparinase by Bacteroides. *J. Bact.,* **81**, 595.

Gibbons, R. J. & MacDonald, J. B. (1960) Haemin and Vitamin K compounds as required factors for the cultivation of certain strains of Bacteroides melaninogenicus. *J. Bact.,* **80**, 164.

Gibbons, R. J., Socransky, S. S., Sawyer, S., Kapsimalis, B. & MacDonald, J. B. (1963) The microbiota of the gingival crevice area of man. ii the predominant cultivable organisms. *Archs. oral. Biol.,* **8**, 281.

Gillespie, W. A. & Guy, J. (1956) Bacteroides in intra-abdominal sepsis their sensitivity to antibiotics. *Lancet,* **1**, 1039.

Gorbach, S. L., Menda, K. B., Thadepalli, H. & Keith, L. (1973) Anaerobic microflora of the cervix in healthy women. *Am. J. Obstet. Gynec.,* **117**, 1053.

Gorbach, S. L. & Bartlett, J. G. (1974) Anaerobic infections. Three parts. *New Eng. J. Med.,* **290**, p. 1177, 1237, 1289.

Gorbach, S. L. & Thadepalli, H. (1974) Clindamycin in pure and mixed anaerobic infections. *Arch. intern. Med.,* **134**, 87.

Gorbach, S. L., Thadepalli, H. & Norsen, J. (1974) Anaerobic microorganisms in intra-abdominal infections. *In: Anaerobic bacteria; their role in disease. A Symposium.* Springfield, Illinois: Charles C. Thomas.

Griffin, M. H. (1970) Fluorescent antibody techniques in the identification of the Gram negative non-spore forming anaerobes. *Hlth. Lab. Sci.,* **7**, 78.

Gunn, A. A. (1956) Bacteroides septicaemia. *Jl. R. Coll. Surg. Edinb.,* **2**, 41.

Haenel, H. (1961) Quoted by Drasar, B. S. (1967)

Haldane, E. V. & van Rooyen, C. E. (1972) Treatment of severe bacteroides infections with parenteral clindamycin. *Can. med. Ass. J.,* **107**, 1177.

Hall, W. L., Sobel, A. I., Jones, C. P. & Parker, R. T. (1967) Anaerobic post-operative pelvic infections. *Obstet. Gynec., N.Y.,* **30**, 1.

Hallé, J. (1898) Recherches sur la Bactériologie du Canal Génital de la Femme. Thèse de Paris.

Harris, J. W. & Brown, J. H. (1927) Description of a new organism that may be a factor in the causation of puerperal infections. *Bull. Johns Hopkins Hosp.,* **40**, 203.

Hawbaker, E. L. (1971) Thyroid abscess. *Am. Surg.,* **37**, 290.

Headington, J. T. & Beyerlein, B. (1966) Anaerobic bacteria in routine urine culture. *J. clin. Path.,* **19**, 573.

Heineman, H. S. & Braude, A. I. (1963) Anaerobic infection of the brain. *Am. J. Med.,* **34**, 682.

Heller, C. L. (1954) A simple method or producing anaerobiosis. *J. appl. Bact.,* **17**, 202.

Hentges, D. J. (1970) Enteric pathogen-normal flora interactions. *Am. J. clin. Nutr.,* **23**, 1451.

Hermans, P. E. & Washington, J. A. (1970) Polymicrobial bacteraemia. *Ann. intern. Med.,* **73**, 387.

Hine, M. K. & Berry, G. P. (1937) Quoted by Gillespie & Guy.

Hite, K. E., Hesseltine, H. C. & Goldstein, L. (1947) A study of the bacterial flora of the normal and pathologic vagina and uterus. *Am. J. Obstet. Gynec.,* **53**, 233.

Hoffmann, K. & Gierhake, F. W. (1969) Postoperative infection of wounds by anaerobes. *Germ. med. Mon.,* **14**, 31.

Hofstad, T. & Kristoffersen, T. (1970) Chemical characteristics of endotoxin from Bacteroides fragilis. NCTC 9343. *J. gen. Microbiol.,* **61**, 15.

Holt, R. J. & Stewart, G. T. (1964) Production of amidase and lactamase by bacteria. *J. gen. Microbiol.,* **36**, 203.

van Houte, J. & Gibbons, R. J. (1966) Studies of the cultivable flora of normal human faeces. *Antonie van Leeuwenhoek,* **32**, 212.

Hubbert, W. T. & Rosen, M. N. (1970) Pasteurella multocida infection due to animal bite. *Am. J. publ. Hlth,* **60**, 1103.

Hungate, R. E. (1950) The anaerobic mesophilic cellulolytic bacteria *Bact. Rev.,* **14**, 1.

Ingham, H. R., Selkon, J. B., Codd, A. A. & Hale, J. H. (1968) A study in vitro of the sensitivity to antibiotics of Bacteroides fragilis. *J. clin. Path.,* **21,** 432.

Ingham, H. R., Selkon, J. B., Codd, A. A. & Hale, J. H. (1970) The effect of carbon dioxide on the sensitivity of Bacteroides fragilis to certain antibiotics in vitro. *J. clin. Path.,* **23,** 254.

Janecka, I. P. & Rankow, R. M. (1971) Fatal mediastinitis following retropharyngeal abscess. *Archs. Otolar.,* **93.** 630.

Kagnoff, M. F., Armstrong, D. & Blevins, A. (1972) Bacteroides bacteraemia experiences in a hospital for neoplastic diseases. *Cancer. N.Y.,* **29,** 245.

Kane, P. O., McFadzean, J. A. & Squires, S. (1961) Absorption and excretion of metronidazole. Part 1. Serum concentration and urinary excretion after oral administration. *Br. J. vener. Dis.,* **37,** 273.

Kelly, P. J., Martin, W. J. & Coventry, M. B. (1970) Bacterial (suppurative) arthritis in the adult. *J. Bone, Jt. Surg.,* **52,** 1595.

Khairat, O. (1964) Modernising Brewer and other anaerobic jars. *J. Bact.,* **87,** 963.

Killgore, G. E., Starr, S. E., del Bene, V. E., Whaley, D. N. & Dowell, V. R. (1973) Comparison of 3 anaerobic systems for the isolation of anaerobic bacteria from clinical specimens. *Am. J. clin. Path.,* **59,** 552.

Kislak, J. W. (1972) The susceptibility of Bacteroides fragilis to 24 antibiotics. *J. infect. Dis.,* **125,** 295.

Kissling, K. (1929) Quoted by Gunn, A. A.

Kline, B. S. & Berger, S. S. (1935) Pulmonary abscess and pulmonary gangrene; analysis of 90 cases observed in 10 years. *Archs. intern. Med.,* **56,** 753.

Kuklinca, A. G. & Gavan, T. L. (1969) The culture of sterile urine for detection of anaerobic bacteria—not necessary for standard evaluation. *Cleveland. Clin. Q.,* **36,** 133.

Leake, D. L. (1972) Bacteroides osteomyelitis of the mandible. A report of two cases. *Oral. Surg.,* **34,** 585.

Ledger, W. J., Campbell, C. & Willson, J. R. (1968) Postoperative adnexal infections. *Obstet. Gynec. N.Y.,* **31,** 83.

Ledger, W. J., Sweet, R. L. & Headington, J. T. (1971) Bacteroides species as a cause of severe infections in Obstetric and gynaecologic patients. *Surg. Gynec. Obstet.,* **133,** 837.

Ledger, W. J., Kriewall, T. J., Sweet, R. L. & Fekety, F. R. (1974) The use of parenteral clindamycin in the treatment of Obstetric Gynaecologic patients with severe infections. *Obstet. Gynec. N.Y.,* **43,** 490.

Lee, T. H. & Berg, R. B. (1971) Cephalhaematoma infected with Bacteroides. *Am. J. Dis. Child.,* **121,** 77.

Leigh, D. A. (1974) Clinical importance of infections due to *Bacteroides fragilis* and role of antibiotic therapy. *Br. med. J.,* **3,** 225.

Leigh, D. A., Simmons, K. & Norman, E. (1974) Bacterial flora of the appendix fossa in appendicitis and postoperative wound infections. *J. clin. Path.,* **27.** 997.

Leigh, D. A., Kershaw, M. & Simmons, K. (1976) Vaginal flora related to various groups of women. *Scot. Med. J.,* **20,** 228.

Leigh, D. A. & Simmons, K. (1975a) Activity of cephazolin and other cephalosporins against *Bacteroides fragilis.* To be published.

Leigh, D. A. & Simmons, K. (1975b) Activity of Minocycline against *Bacteroides fragilis. Lancet.,* **1,** 51.

Leigh, D. A. (1975) Activity of spectinomycin against strains of *Bacteroides fragilis*. Unpublished observations.

Lemierre, A. (1936) On certain septicaemias due to anaerobic organisms. *Lancet*, 1, 701.

Lewis, C., Clapp, H. W. & Grady, J. E. (1963) In vitro and in vivo evaluation of lincomycin, a new antibiotic. *Antimicrob. Ag. Chemother*, 570.

Lewis, J. F. (1973) Myocardial infarction during pregnancy with associated myocardial Bacteroides abscess. *Sth. med. J., Nashville*, 66, 379.

Lewy, B. (1891). Quoted by Gunn, A. A.

Lifshitz, F., Liu, C. & Thurn, A. N. (1963) Bacteroides meningitis. *Am. J. Dis. Child.*, 105, 487.

Lindell, T. D., Fletcher, W. S. & Krippaehne, W. W. (1973) Anorectal suppurative disease, a retrospective review. *Am. J. Surg.*, 125, 189.

Lodenkamper, H. & Steinen, G. (1955) Importance and therapy of anaerobic infections. *Antibiotic. Med. clin. Ther.*, 1, 653.

Loeffler, F. (1884) Quoted by Gillespie, W. A. & Guy, J.

Loesche, W. J. & Gibbons, R. J. (1965) A practical scheme for identification of the most numerous oral gram negative anaerobic rods. *Archs. oral. Biol.*, 10, 723.

Lukasik, J. (1963) A comparative evaluation of the bacteriological flora of the uterine cervix and fallopian tubes in cases of salpingitis. *Am. J. Obstet. Gynec.*, 87, 1028.

MacDonald, J. B., Socransky, S. S. & Gibbons, R. J. (1963) Aspects of the pathogenesis of mixed anaerobic infections of mucous membranes. *J. dent. Res.*, 42, 529.

Mackenzie, I. & Litton, A. (1974) Bacteroides bacteraemia in surgical patients. *Br. J. Surg.*, 61, 288.

Maier, R. L. (1937) Human bite infections of the hand. *Ann. Surg.*, 106, 423.

Mallory, A., Savage, D., Kern, F. & Smith, J. G. (1973) Patterns of bile acids and microflora in the human small intestine. ii., Microflora. *Gastroenterology* 64, 34.

Marcoux, J. A., Zabransky, R. J., Washington, J. A., Wellman, W. E. & Martin, W. J. (1970) Bacteroides bacteraemia. *Minn. Med.*, 53, 1169.

Martin, W. J., Gardner, M. & Washington, J. A. (1972) In vitro antimicrobial susceptibility of anaerobic bacteria isolated from clinical specimens. *Antimicrob. Ag. Chemother.*, 1, 148.

Masri, A. F. & Grieco, M. H. (1972) Bacteroides endocarditis, report of a case. *Am. J. med. Sci.*, 263, 357.

McHenry, M. C., Wellman, W. E. & Martin, W. J. (1961) Bacteraemia due to *Archs. intern. Med.*, 107, 572.

McLaughlin, P., Meban, S. & Thompson, W. G. (1973) Anaerobic liver abscess complicating radiation enteritis. *Can. med. Ass. J.*, 108, 353.

McVay, L. V. & Sprunt, D. H. (1952) Bacteroides infections. *Ann. intern. Med.*, 36, 56.

Meleney, F. L. (1949) *Clinical aspects and treatment of surgical infections.* Philadelphia: W. B. Saunders Co.

Miller, C. P. & Bornhoff, M. (1963) Changes in the mouse's enteric flora associated with enhanced susceptibility to Salmonella infection following streptomycin treatment. *J. infect. Dis.*, 113, 59.

Mitchell, A. A. B. (1973) Incidence and isolation of Bacteroides species from clinical material and their sensitivity to antibiotics. *J. clin. Path.*. **26**, 738.

Mitchell, A. A. B. & Simpson, R. G. (1973) Bacteroides septicaemia. *Curr. Med. Res. Opinion*, **1**, 385.

Moore, W. E. C. (1966) Techniques for routine culture of fastidious anaerobes. *Int. J. Systemat. Bacteriol.*, **16**, 173.

Nastro, L. J. & Finegold, S. M. (1972) Bactericidal activity of 5 antimicrobial agents against Bacteroides fragilis. *J. infect. Dis.*, **126**, 104.

Nastro, L. J. & Finegold, S. M. (1973) Endocarditis due to anaerobic gram negative bacilli. *Am. J. med. Sci.*, **54**, 482.

Neary, M. P., Allen, J., Okubadejo, O. A. & Payne, D. J. H. (1973) Preoperative vaginal bacteria and postoperative infections in gynaecological patients. *Lancet*, **21**, 1291.

Nelson, J. D. & Koontz, W. C. (1966) Septic arthritis in infants and children, a review of 117 cases. *Pediatrics.*, *Springfield*, **38**, 966.

Nettles, J. L., Kelly, P. J., Martin, W. J. & Washington, J. A. (1969) Musculo-skeletal infections due to Bacteroides. *J. bone Joint Surg.*, **51**, 601.

Ninomiya, K., Ohtani, F., Koosaka, S., Kamiya, H., Ueno, K., Susuki, S. & Inque, T. (1972) Simple and expedient methods for differentiation among Bacteroides, Sphaerophorus and Fusobacterium. *Jap. J. med. Sci. Biol.*, **25**, 63.

Nissle, A. (1916) Über die Gründlagen einer insächlichen Bekämpfung der pathologischen Darmflora. *Dt. med. Wschr.*, **42**, 1181.

Nobles, E. R. (1973) Bacteroides infections. *Ann. Surg.*, **177**, 601.

Norris, C. (1901) Quoted by Gunn, A. A. q.v.

Okubadejo, O. A., Green, P. J. & Payne, D. J. H. (1973) Bacteroides infections among hospital patients. *Br. med. J.*, **2**, 212.

Okubadejo, O. A., Green, P. J. & Payne, D. J. H. (1974) Bacteroides in the blood. *Br. med. J.*, **1**, 147.

Parker, R. T. & Jones, C. P. (1966) Anaerobic pelvic infections and develop-ments in hyperbaric oxygen therapy. *Am. J. Obstet. Gynec.*, **96**, 645.

Pasteur, L. (1861–63) Quoted by Gunn, A. A., & Hillas-Smith, in Anaerobic infections, a review (1974) *Curr. med. Res. Opinion*, **2**, 109.

Pearson, H. E. (1967) Bacteroides in areolar breast abscesses. *Surg. Gynec. Obstet.*, **125**, 800.

Pearson, H. E. & Anderson, G. V. (1967) Perinatal deaths associated with bacteroides infections. *Obstet. Gynec.*.

Pearson, H. E. & Smiley, D. F. (1968) Bacteroides in pilonidal sinuses. *Am. J. Surg.*, **115**, 336.

Pearson, H. E. & Anderson, G. V. (1970a) Genital bacteroidal abscesses in women. *Am. J. Obstet. Gynec.*, **107**, 1264.

Pearson, H. E. & Anderson, G. V. (1970b) Bacteroides infections and preg-nancy. *Obstet. Gynec., N.Y.*, **35**, 31.

Pearson, H. E. & Harvey, J. P. (1971) Bacteroides infections in orthopaedic conditions. *Surg. Gynec. Obstet.*, **132**, 876.

Percival, A. & Roberts, C. (1972) Isolation of Bacteroides from clinical specimens. *J. med. Microbiol.*, **5**, 13.

Phillips, I. & Warren, C. (1975) Susceptibility of Bacteroides fragilis to spectinomycin. *J. Antimicrobial. Chemother.*, **1**, 91.

Pinkus, G., Veto, G. & Braude, A. I. (1968) Bacteroides penicillinase. *J. Bact.*, **96**, 1437.

Post, F. J., Allen, A. D. & Reid, T. C. (1967) Simple medium for the selective isolation of Bacteroides and related organisms and their occurrence in sewage. *Appl. Microbiol.*, **15**, 213.

Prévot, A. R. (1938) Études de systématique bactérienne invalidité du genre Bacteroides, Remembrement et reclassification. *Anns. Inst. Pasteur*, **60**, 285.

Prévot, A. R. (1940) Chimiothérapie des septicémies expérimentales du lapin à spherophorus funduliformis. *C.r. Séanc. Soc. Biol.*, **134**, 90.

Prévot, A. R. (1948) *Manuel de classification et de determination des bactéries anaérobies*, p. 290, 2nd ed. Paris: Masson.

Prévot, A. R. (1957) *Manuel de classification et de détermination des bactéries anaérobies*, p. 362, 3rd ed. Paris: Masson.

Prévot, A. R. (1966) *Techniques pour la diagnostic des bactéries anaérobies*, p. 153, 2nd ed. St-Maude, France: la Tourelle.

Prince, H. N., Grunberg, E., Titsworth, E. & Howe, J. S. (1969) Effects of 1-(21nitro-1-imidazolyl) 3-methoxy-2-propanol and 2-methyl-5-nitroimidazole-1-ethanol against anaerobic and aerobic bacteria and protozoa. *Appl. Microbiol.*, **18**, 728.

Quinto, G. (1964) Identification of non-sporulating anaerobes. *Am. J. med. Technol.*, **30**, 304.

Reid, J. D., Snider, G. E., Toone, E. C. & Howe, J. S. (1945) Anaerobic septicaemia: report of six cases with clinical, bacteriologic and pathologic studies. *Am. J. med. Sci.*, **209**, 296.

Rein, J. M. & Cosman, B. (1971) Bacteroides necrotising fasciitis of the upper extremity. *Plastic reconstr., Surg.*, **48**, 592.

Rendu, M. & Rist, M. E. (1899) Étude clinique et bactériologique de trois cas de pleurésis putride. *Bull. Mém. Soc. méd. Hôp. Paris*, **16**, 133.

Rist, E. (1898) Études bactériologiques sur les infections d'origine otique. *Thès. Fac. Méd. Paris, 1898*, p. 175.

Romond, C., Beerens, H. & Wattre, P. (1972) Identification sérologique des Bactéroides en relation avec leur poirron pathogène. *Archs. roum. Path. exp. Microbiol.*, **31**, 351.

Rosebury, T. (1962) *Microorganisms indigenous to Man*. p. 435, New York: McGraw-Hill.

Rosebury, T. & Reynolds, J. B. (1964) Continuous anaerobiosis for cultivation of spirochaetes. *Proc. Soc. exp. biol. Med.*, **117**, 813.

Rosenblatt, J. E., Fallon, A. M. & Finegold, S. M. (1973) Comparison of methods for isolation of anaerobes from clinical specimens. *Appl. Microbiol.*, **25**, 77.

Rotheram, E. B. & Schick, S. F. (1969) Non clostridial anaerobic bacteria in septic abortion. *Am. J. Med.*, **46**, 80.

Rubin, S. H. & Boyd, L. J. (1958) Clinical aspects of bacteroides infections in gastroenterology. *Am. J. Gastroent.*, **29**, 131.

Sabbaj, J., Sutter, V. L. & Finegold, S. M. (1972) Anaerobic pyogenic liver abscess. *Ann. intern. Med.*, **77**, 629.

Saksena, D. S., Block, M. A., McHenry, M. C. & Truant, J. P. (1968) Bacteroidaciae: anaerobic organisms encountered in surgical infection. *Surgery, St. Louis.* **63**, 261.

Salem, A. R., Jackson, D. D. & McFadzean, J. A. (1975). An investigation of interactions between metronidazole (Flagyl) and other antibacterial agents. *J. Antimicrob. Chemother.*, **1**, 387.

Salibi, B. S. (1964) Bacteroides infection of the brain. *Archs. Neurol.,* 10, 629.

Schatten, W. E. (1956) Intraperitoneal antibiotic administration in the treatment of acute bacterial peritonitis. *Surg. Gynaec. Obstet.,* 102, 339.

Schmitz, H. (1930) Quoted by Gillespie, W. A. and Guy, J. (1956).

Schoolman, A., Liu, C. & Rodecker, C. (1966) Brain abscess caused by bacteroides infection. *Archs. intern. Med.,* 118, 150.

Schoutens, E., Labbé, M. & Yourassowsky, E. (1973) Septicémies a Bacteroides fragilis, incidence et sensibilité des souches aux antibiotiques. *Path. Biol.,* 21, 349.

Schumer, W., Nichols, R. L., Miller, B., Samer, E. T. & MacDonald, G. O. (1973) Clindamycin in the treatment of soft tissue infections. *Archs. Surg., Chicago,* 106, 578.

Schwartz, O. H. & Dieckmann, W. J. (1927) Puerperal infection due to anaerobic streptococci. *Am. J. Obstet. Gynec.,* 13, 467.

Sebald, M. (1962). Étude sur les bactéries anaérobies gram négatives asporulées. Thès. Sci. Barnéoud Laval, France, p. 161.

Selkon, J. B., Hale, J. H. & Ingham, H. R. (1975) Metronidazole in the treatment of anaerobic infection in man. *Proc. Internat. Congr. Chemother., Lond.* To be published.

Sen, P., Tecson, F., Kapila, R. & Louria, M. D. (1974) Clindamycin in the oral treatment of putative anaerobic pneumonias. *Arch. intern. Med.,* 134, 73.

Shimada, K., Sutter, V. L. & Finegold, S. M. (1970) Effect of bile and deoxycholate on gram negative anaerobic bacteria. *Appl. Microbiol.,* 20, 737.

Shinn, D. L. S. (1962) Metronidazle in acute ulcerative gingivitis. *Lancet,*1, 1191.

Shinn, D. L. S., Squires, S. & McFadzean, J. A. (1965) The treatment of Vincent's disease with metronidazole. *Dent. Pract.,* 15, 275.

Sinkovics, J. G. & Smith, J. P. (1970) Septicaemia with bacteroides in patients with malignant disease. *Cancer, N.Y.,* 25, 663.

Slanetz, L. W. & Rettger, L. F. (1933) A systematic study of the fusiform bacteria. *J. Bact.,* 26, 599.

Smith, C. D., Deane, C., Montgomery, J., Williams, R. & Sampson, C. C. (1968) Bacteroides, incidence and studies on 100 isolations. *J. nat. med. Ass.,* 60, 215.

Smith, D. T. (1930) Fuso-spirochaetal disease of the lungs produced with cultures from Vincent's angina. *J. infect. Dis.,* 46, 303.

Smith, D. T. (1932) *Oral spirochaetes and related organisms in fusospirochaetal diseases.* Baltimore: Williams and Wilkins, Co.

Socransky, S. S., MacDonald, J. B. & Sawyer, S. (1959) The cultivation of Treponema microdentium as surface colonies. *Arch. oral. Biol.,* 1, 171.

Socransky, S. S. & Gibbons, R. J. (1965) Required role of Bacteroides melaninogenicus in mixed anaerobic infections. *J. infect. Dis.,* 115, 247.

Sonnenwirth, A. C. (1960) *A study of certain gram negative non sporulating anaerobic bacteria indigenous to man with special reference to their classification by serological means.* A Dissertation. p. 130, St Louis, Missouri: Washington University.

Sonnenwirth, A. C., Yin, E. T., Sarmiento, E. M. & Wessler, S. (1972) Bacteroidaciae endotoxin detection by Limulus assay. *Am. J. clin. Nutr.,* 25, 1452.

Sonnenwirth, A. C. (1974) Incidence of intestinal anaerobes in blood cultures. *In: Anaerobic bacteria, their role in disease,* Symposium. Springfield,

Illinois: Charles C. Thomas.

Spiers, M. (1971) Classification systems of the Bacteroides group. *Med. Lab. Tech.*, **28**, 360.

Statmen, N. & Spitzer, S. (1962) Bacteroides septicaemia and osteomyelitis. *Ohio Med. J.*, **58**, 1374.

Stokes, E. J. (1958) Anaerobes in routine diagnostic cultures. *Lancet*, **1**, 668.

Sutter, V. L., Sugihara, P. T. & Finegold, S. M. (1971) Rifampin-Blood Agar as a selective medium for the isolation of certain anaerobic bacteria. *Appl. Microbiol.*, **22**, 777.

Sutter, V. L. & Finegold, S. M. (1971) Antibiotic disc susceptibility test for rapid presumptive identification of gram negative anaerobic bacteria. *Appl. Microbiol.*, **21**, 13.

Sutter, V. L., Kwok, Y. & Finegold, S. M. (1972) Standardised antimicrobial disc susceptibility testing of anaerobic bacteria. 1. Susceptibility of Bacteroides fragilis to tetracycline. *Appl. Microbiol.*, **23**, 268.

Sukuzi, S., Ushijima, T. & Ichinose, H. (1966) Differentiation of Bacteroides from Sphaerophorus and Fusobacterium. *Jap. J. Microbiol.*, **10**, 193.

Swenson, R. M., Michaelson, T. C., Daly, M. J., Faco, G. & Spaulding, E. H. (1973) Anaerobic bacterial infections of the female genital tract. *Obstet. Gynec., N.Y.*, **42**, 538.

Swenson, R. M. & Lorber, B. (1974) Bacteriology of aspiration pneumonia community vs hospital acquired. *Clin. Res.*, **22**, 380.

Swartz, M. N. & Karchmer, A. W. (1974) Anaerobic infections of the central nervous system. *In: Anaerobic bacteria, their role in disease*, Symposium. Springfield, Illinois: Charles C. Thomas.

Thadepalli, H., Gorbach, S. L. & Keith, L. (1973a) Anaerobic infections of the female genital tract: bacteriologic and therapeutic aspects. *Am. J. Obstet. Gynec.*, **117**, 1034.

Thadepalli, H., Gorbach, S. L. & Bartlett, J. G. (1973b) Apparent failure of chloramphenicol in anaerobic infections. *13th Interscience Conference on Antimicrobial Agents and Chemotherapy*: Washington.

Thompson, L. & Beaver, D. C. (1932) Bacteraemias due to anaerobic gram negative organisms of the genus Bacteroides. *Med. Clins. N. Am.*, **15**, 1611.

Tiessier, P., Reilly, J., Rivalier, E. & Stefanesco, V. (1931) Les septicémies primitive dues au Bacillus funduliformis. Étude, clinique, bacteriologique et expérimentale. *Ann. Med.*, **30**, 97.

Tracy, O., Gordon, A. W., Moran, F., Love, W. C. & McKenzie, P. (1972) Lincomycins in the treatment of Bacteroides infections. *Br. med. J.*, **1**, 280.

Tumulty, P. A. & McGehee, H. A. (1948) Experiences in the management of subacute bacterial endocarditis treated with penicillin. *Am. J. med. Sci.*, **47**, 37.

Tunnicliff, R. (1906) Quoted by Gillespie, W. A. & Guy, J.

Tynes, B. S. & Frommeyer, W. J. (1962) Bacteroides septicaemia. *Ann. intern. Med.*, **56**, 12.

Urdal, K. & Berdal, P. (1949) The microbial flora in 81 cases of maxillary sinusitis. *Acta. oto-lar.*, **37**, 20.

Veillon, A. & Zuber, A. (1898) Recherches sur quelques microbes strictement anaérobies et leur rôle en pathologie. *Archs. Méd. exp. Anat. path.*, **10**, 517.

Vincent, M. H. (1896) Sur l'étiologie et sur le lésions anatomopathologique de la pourriture d'hôpital. *Annls. Inst. Pasteur*, **10**, 488.

Vinke, B. & Borghans, J. G. A. (1963) Bacteroides as a cause of suppuration and septicaemia. *Trop. geogr. Med.*, **15**, 76.

Watt, B. (1973) The influence of carbon dioxide on the growth of obligate and facultative anaerobes on solid media. *J. med. Microbiol.*, **6**, 307.

Weinberg, M., Prévot, A. R., Davesne, J. & Renard, C. (1–28) Recherches sur la bactériologie et la serothérapie des appendicites argues. *Annls. Inst. Pasteur*, **42**, 1167.

Weinstein, L. & Rubin, R. H. (1973) infective endocarditis. *Prog. cardiovasc. Dis.*, **16**, 239.

Weiss, J. E., James, E. & Rettger, L. F. (1937) The gramnegative bacteroides of the intestine. *J. Bact.*, **33**, 423.

Werner, H. (1965) Zum gegenwärtigen Stand der Diagnostik der Gram-negativen anaeroben sporenlosen Stabchen. *Zentbl. Bakt. ParasitKde.*, **197**, 407.

Werner, H. (1972) The susceptibility of Bacteroides, Fusobacterium, Leptotrichia, and Sphaerophorus strains to rifamicin. *Arzneimittel-Forsch.*, **22**, 1043.

Whelan, J. P. F. & Hale, J. H. (1973) Bactericidal activity of metronidazole against *Bacteroides fragilis*. *J. clin. Path.*, **26**, 393.

van der Wiel-Korstanje, J. A. A. & Winkler, K. C. (1970) Medium for the differential count of the anaerobic flora in human faeces. *Appl. Microbiol.*, **20**, 168.

Wierdsma, J. G. & Clayton, E. M. (1964) The effects of certain antibiotics on the normal postpartum intrauterine bacteriologic flora. *Am. J. Obstet. Gynec.*, **88**, 541.

Willis, A. T. & Study Group (1974) Metronidazole in the prevention and treatment of Bacteroides infections in gynaecological patients. *Lancet*, **2**, 1540.

Wilson, W. R., Martin, W. J., Wilkowske, C. J. & Washington, J. A. (1972) Anaerobic bacteraemia. *Mayo. Clin. Proc.*, **47**, 639.

Zabransky, R. J. (1970) Isolation of anaerobic bacteria from clinical specimens. *Mayo. Clin. Proc.*, **45**, 256.

Ziment, I., Davis, A. & Finegold, S. M. (1969) Joint infection by anaerobic bacteria. A case report and review of the literature. *Arthritis and Rheumatism*, **12**, 627.

# 6 | The Viridans Streptococci
## G. COLMAN

*The viridans group has regained the respectability it lost in the hey-day of serology*—Cowan (1974).

## Introduction

The viridans streptococci are of current interest, perhaps, because one species, *Streptococcus mutans*, seems to have a special relationship to the initiation of dental caries. Order was brought to the pyogenic streptococci, the enterococci and the lactic streptococci more than thirty years ago, but some species in the viridans streptococci have been confirmed only within the last few years. In the past one heterogeneous species, *Str. mitis*, was defined in terms of the absence of recognizable characteristics. This tradition is continued in the eighth edition of Bergey's Manual (Buchanan & Gibbons, 1974). An intuitively satisfactory bacterial species is however formed when the strains allotted to the species have a number of characteristics in common, as exemplified by *Str. agalactiae*—Lancefield group B (Munch–Petersen, 1954).

For a medical microbiologist the viridans streptococci would include *Str. mutans, Str. bovis, Str. sanguis, Str. salivarius, Str. milleri, Str. mitior, Str. pneumoniae* and a number of strains which cannot yet be placed in satisfactory species (See Table 6.1). It is immediately apparent from this list that some cultures that would properly be classed in the species listed do not produce greening when grown on blood agar. Some do not produce any apparent change in the blood, the non-haemolytic streptococci, fewer are apparently haemolytic but do not produce soluble haemolysins.

## *Streptococcus mutans*

*Streptococcus mutans* was the name proposed by Clarke (1924) for the organism that he isolated from the softened dentine of dental

**Table 6.1.** *Some properties of the viridans streptococci of man* * +81 −100 per cent of*

|  | *Str. mutans* | *Str. bovis* | *Str. sanguis* | *Str. sanguis*† |
|---|---|---|---|---|
| Changes in blood agar | indifferent, may be some greening or haemolysis | indifferent, may be some greening | greening or haemolysis | greening |
| Sucrose broth | dextran, adherent | dextran, gel | dextran, gel | dextran, gel |
| Lancefield group reactions | few E | D | some H | — |
| Hydrolysis of: |  |  |  |  |
|   arginine | − | − | + | − |
|   aesculin | + | + | d | − |
|   starch | − | + | −‖ | − |
| Hydrogen peroxide formed | − | − | + | + |
| Fermentation of: |  |  |  |  |
|   mannitol | + | + | − | − |
|   sorbitol | + | − | − | − |
| Bile soluble *and* optochin sensitive | − | − | − | − |

\* None grow in 6·5 per cent NaCl broth or hydrolyse hippurate.
† These strains are the dextran-forming variety of *Str. mitior.*
‡ If dextran production is ignored some of these strains cannot be distinguished from *Str. sanguis.*

*strains positive; d 21–80 per cent strains positive; — 0–20 per cent of strains positive.*

| unclassified‡ | Str. salivarius | Str. milleri | Str. mitior | Str. pneumoniae |
|---|---|---|---|---|
| usually greening | usually indifferent | indifferent, haemolysis or slight greening | greening or haemolysis | greening |
| — | levan, gel | — | — | — |
| some H or K | some K | some F, C, G or A | some O, K or M | — |
| + | — | + | — | — |
| d | + | + § | — | — |
| —‖ | — | — | —‖ | —‖ |
| + | — | — | + | + |
| — | — | — | — | — |
| — | — | — | — | — |
| — | — | — | — | + |

§ Notable exceptions occur and are discussed in the text
‖ If starch hydrolysis is tested in an agar medium some strains will show narrow zones of clearing.

caries in man. It is unlikely that this was the first description of this species. Heim's description (1924) of *Str. halitus* predates that given by Clarke and the source and manner of growth given for *Str. halitus* suggests that Heim and Clarke were describing the same organism. Abercrombie and Scott (1928) found *Str. mutans* to be a cause of subacute bacterial endocarditis and interest in the species was maintained for a few years among those working with oral streptococci (Maclean, 1927; Crowe, 1927; Thomson & Thomson, 1929). Recent interest in this micro-organism dates from the experiments in gnotobiotic rats in which an un-named streptococcus was most efficient in causing dental caries (Orland et al., 1955). Similar strains were isolated from hamsters (Keyes, 1960) and the presence of this organism in the oral flora of these rodents was one of the necessary conditions for the development of dental caries. In man comparable strains of *Str. mutans* are found in increased numbers on those surfaces of the teeth that have become carious (de Stoppelaar et al., 1969). Other bacteria, *Actinomyces viscosus* for example, can cause dental caries in experimental animals (Llory et al., 1971), but the part played by these bacteria in the human disease is not yet clear.

*Str. mutans*, when present in the human oral cavity, colonizes the surfaces of teeth (Gibbons, 1972) and can be isolated only with difficulty from other parts of the mouth. A selective medium for this species has been described (Gold et al., 1973) based on Mitis Salivarius Agar which is made even more selective by increasing the sucrose content of the medium and by adding bacitracin. A second medium, one that does not contain antibiotics, such as the 'TYC' medium described by de Stoppelaar et al. (1967), can usefully be used alongside the bacitracin medium because some strains of *Str. mutans* are sensitive to the bacitracin in Gold's medium and to the sulphonamide used by Carlsson (1967a) for the same purpose. Cultures of *Str. mutans* should be incubated in an atmosphere containing a supplement of carbon dioxide. The colonial form of this organism, while recognizable for a given strain, shows such marked differences between strains that physiological or serological tests are necessary for the identification of freshly isolated cultures. Many strains are non-haemolytic when grown on horse blood agar but colonies of some strains are surrounded by narrow zones of greening.

The characteristics of the species have recently been the subject of two review articles (Perch et al., 1974; Facklam, 1974). In outline these streptococci ferment mannitol and sorbitol, form dextran, hydrolyse aesculin, yield acetoin from glucose, will grow in 4 per cent sodium chloride broth, but not in 6·5 per cent, fail to release ammonia by hydrolysis of arginine, do not produce hydrogen peroxide during growth, do not grow at 45°C (113°F) or hydrolyse starch (Colman & Williams, 1972, 1973). There are exceptions to this combination of characteristics. The strains present in the fissures of rat molars, strains of serotype b, hydrolyse arginine and may grow at 45°C (113°F). Cultures of the serotypes d–g may neither ferment sorbitol, nor hydrolyse aesculin but do form peroxide during growth (Perch et al., 1974).

Seven serotypes of *Str. mutans* have now been described (Perch et al., 1974) and one serotype cross-reacts with Lancefield group E. If sera are to be used only for identification then nearly all of the strains isolated from humans can be identified by the Lancefield precipitin technique using two unabsorbed sera. Suitable antisera are those prepared against a type c strain. NCTC 10449 for instance, and the second against a strain such as OMZ65. The antisera against type c strains will usually cross-react with extracts from strains of type e, and antisera to the strain OMZ65 generally yield precipitates with serotypes a, d and g. (Perch et al., 1974). Type c is the serotype of *Str. mutans* most frequently isolated from cases of subacute bacterial endocarditis. In a collection of streptococci isolated from 222 cases of endocarditis, or suspected endocarditis, 13 per cent were identified as *Str. mutans* (Parker & Ball, 1975). Strains of this species isolated from blood are sensitive to penicillin (Facklam, 1974).

*Str. mutans* forms dextran when grown in the presence of sucrose. This property cannot usually be deduced from the colonial form on sucrose agar, but dextran can be demonstrated by precipitation from sucrose broth either by ethanol or by type II pneumococcus antiserum (Hehre & Neill, 1946). Strains of *Str. mutans* also produce dextranases (Staat & Schachtele, 1974) that are capable of breaking down the 1:6 links in dextran chains (Guggenheim, 1975). The branched dextrans that remain in culture fluids (Ebisu et al., 1974) are presumably the net result of the action of the glucosyltransferases and dextranases produced by the one organism.

## Streptococcus bovis

*Str. bovis* is recognized as one of the species of viridans streptococci associated with subacute bacterial endocarditis (Niven et al., 1948). Endocarditis caused by *Str. bovis* appears to be restricted to the middle-aged and elderly (Parker & Ball, 1975). The natural habitat of this organism in man is unknown.

Human strains of *Str. bovis* (Lancefield group D) ferment mannitol, lactose and sucrose, grow on 40 per cent bile agar and hydrolyse aesculin, grow at 45°C (113°F) but not at 50°C (122°F), hydrolyse starch, usually form dextran from sucrose, but usually do not ferment sorbitol or produce hydrogen peroxide during growth and do not hydrolyse arginine or grow in 6·5 per cent sodium chloride broth.

The Lancefield D antigen is difficult to demonstrate in cultures of *Str. bovis.* The antigen should be precipitated from acid extracts with ethanol (Shattock, 1949) or be prepared by the formamide method from unbuffered broth cultures (Medrek & Barnes, 1962). The backbone of the Lancefield D antigen is glycerol teichoic acid and glycerol teichoic acid is present in strains of *Str. mutans* (Bleiweis et al.. 1971). It is therefore not surprising to find reports of and occasionally observe, cross-reaction between Lancefield D antisera and extracts of *Str. mutans* (de Moor et al., 1968).

The simultaneous demonstration of tolerance to bile and aesculin hydrolysis is possible in one medium (Swan, 1954). These two properties are characteristic of cultures of *Str. faecalis, Str. faecium, Str. uberis, Str. equinus* and *Str. bovis.* Of these five species *Str. faecalis* is the one most commonly isolated from human material. If one therefore relies on the bile-aesculin medium for the presumptive identification of *Str. faecalis* one may miss the occasional strain of *Str. bovis* unless a confirmatory test is used. Among the streptococci, cultures of *Str. faecalis* are notable for their ability to grow in the presence of, and at the same time reduce, the tellurite in Hoyle's medium for the isolation of the diphtheria bacillus. Representatives of *Str. bovis* are inhibited by this concentration of tellurite. They also are more sensitive to antibiotics. Ravreby et al., (1973) readily achieved bactericidal levels of penicillin in the serum of patients with bacterial endocarditis caused by *Str. bovis* and also observed that the infecting strains, again in contrast to *Str. faecalis,* were sensitive to 2 µg/ml lincomycin in broth.

Cultures of *Str. bovis* hydrolyse starch and this is a valuable test for distinguishing this species (Kiel & Skadhauge. 1973). The property can be demonstrated in broth cultures (Raibaud et al.. 1961) or by growing strains on nutrient agar supplemented with soluble starch and demonstrating hydrolysis by flooding the plates with an iodine solution (Andrewes, 1930). Cultures of *Str. bovis* in the latter test show considerable clearing about the growth. Iverson and Millis (1974) have described a medium that allows recognition of starch hydrolysis in growing cultures of *Str. bovis.* Sims (1964) found that strains of *Str. bovis* would grow on acetate agar at pH 5, but only in the presence of carbon dioxide. This property is not found among other species of streptococci, and even *Str. faecalis* is inhibited by acetate agar (Colman, 1967) which is normally used for the selective isolation of lactobacilli. No other medium has apparently been described that is selective for *Str. bovis*. Switzer and Evans (1974) used several commercially available media and found that enterococci were always enriched over *Str. bovis*. A medium demonstrating starch hydrolysis (Iverson and Millis, 1974) and made selective by antibiotics (Colman et al., 1974) might serve for selective isolation of this organism.

When *Str. bovis* is cultured on sucrose agar in the presence of carbon dioxide, large watery colonies are usually formed. In sucrose broth (Bailey & Oxford, 1958a) a diffuse turbidity or gel is seen, unlike the granular adherent growth seen with many strains of *Str. mutans*. The dextran formed from sucrose by *Str. bovis* is a straight chain polymer with 1:6 linkages (Bailey & Oxford, 1958b) unlike the branched polymers formed by glucosyltransferases of *Str. mutans* or *Str. sanguis*. (Cybulska & Pakula, 1963).

Human strains of *Str. bovis* ferment mannitol (Sherman & Stark, 1931; Sherman, 1937) but this property was not found among bovine strains studied by Orla-Jensen (1919). A continuous spectrum apparently exists, on the one hand, between the dextran-forming, starch-hydrolysing strains of *Str. bovis* that are capable of fermenting a variety of carbohydrates and on the other hand, strains described as typical of *Str. equinus* which do not form dextran, have limited ability to hydrolyse starch (Dunican & Seeley, 1962) and ferment few sugars. Both species of course belong to Lancefield group D and, will grow on 40 per cent bile agar but do not hydrolyse arginine or survive 60°C (140°F) for 30 min. Strains intermediate between these two species have been isolated from animal sources (Seeley & Dain, 1960; Mieth, 1962; Smith & Shattock, 1962) and there are indications (Facklam, 1972; Kiel & Skadhauge, 1973) that similar strains might cause human infections.

## Streptococcus sanguis

From the first, *Str. sanguis* has been associated with subacute bacterial endocarditis (Hehre & Neill, 1946; Niven et al., 1946). More recently this organism has been found in large numbers in dental plaque (Carlsson, 1967b) but in smaller numbers in other sites such as human faeces (van Houte et al., 1971). There is however a report (Gledhill & Casida, 1969) that *Str. sanguis* is present in soil.

One of the characteristics of *Str. sanguis* that has attracted attention is the elongation of the cells that can occur in culture. The lengthening occurs at the same time as glycogen accumulates in the cells (Eisenberg et al., 1974) but the prolongation can be prevented by growing the cells anaerobically or in the presence of catalase if cultured in oxygen (Rosan & Eisenberg, 1973).

On blood agar, cultures of *Str. sanguis* produce either greening or haemolysis. The haemolytic strains do not form soluble haemolysins and if the techniques of Brown (1919) are used then apparently alpha-prime haemolysis is seen (Hare, 1935; Gledhill & Casida, 1969). There is general agreement that many strains belong to serological group H, but there is as yet no agreement on a definition of the group H (Colman & Williams, 1972). Rosan (1973) has made an analysis of the antigens of *Str. sanguis* using sera prepared against a single strain. He believes that a negatively charged polymer, which he thinks is a glycerol teichoic acid, is a suitable candidate for the group H antigen. These views are not shared by Bowden and Hardie (1972) who studied material in acid extracts of *Str. sanguis* and found that an antigen reacting with group H sera has a high protein content, contains little, if any, phosphate and has its serological activity destroyed by pronase.

Production of dextran by *Str. sanguis* is a striking property. Other characteristics of the species are the liberation of hydrogen peroxide during growth, the fermentation of lactose and trehalose but not poly-hydric alcohols such as glycerol, mannitol or sorbitol and cultures do not survive 60°C (140°F) for 30 min or grow at 45°C (113°F) or in 6·5 per cent sodium chloride broth. It was however some years (Porterfield, 1950) before it was realized that two dissimilar clusters of dextran-forming strains were being brought together into the one species. The strains of one set usually carry the 'H' antigen or antigens, their cell walls contain rhamnose, they hydrolyse arginine and aesculin and grow on 10 per cent bile agar. Many representatives ferment raffinose, salicin and inulin, and will grow on 40 per cent bile agar. The strains of the second cluster often carry Washburn et al. (1946) type II antigen, their cell walls lack rhamnose, they fail to hydrolyse arginine or aesculin, they are usually inhibited by 10 per cent bile and usually fail to ferment inulin or salicin (Colman & Williams, 1972).

The strains of the second cluster have, if one places on one side their possession of glucosyltransferases, the same characteristics as strains of *Str. mitior* and are probably better classified as the dextran-forming variety of that species. On the other hand there are strains, of which the culture NCTC3165 is an example, that cannot form dextran from sucrose but which in other properties are very similar to *Str. sanguis*. If over-riding importance is given to the ability to form dextran then these last strains must be excluded from the species *Str. sanguis*. If many properties are taken into account in classification then these strains would seem to be an acceptable variety of this species.

## Streptococcus salivarius

Any description of *Streptococcus salivarius* must rest heavily on the work of Sherman and his colleagues. Safford et al. (1937) studied a collection of streptococci isolated from the human throat and found that many fell into one homogeneous species. It seems that because of their interest in extracellular polysaccharides they were sent, by Dr Lancefield, cultures of non-haemolytic streptococci that produced such a polysaccharide from sucrose (Niven et al., 1941a). They found these cultures to be the same as many of the strains they had isolated previously from the throat and named *Str. salivarius* and the polysaccharide was found to be a levan (Niven et al., 1941b).

Sherman, Niven and Smiley's (1943) isolates of *S. salivarius* were defined as non-haemolytic streptococci that formed levan from sucrose and the latter property was highly correlated with a number of other characteristics. Their strains grew at 45°C (113°F) but not at 10°C (50°F) fermented inulin, raffinose and a number of other sugars, hydrolysed aesculin but not arginine and did not ferment mannitol. Later collections showed some differences from these properties. Carlsson's strains (1968) for instance did not grow at 45°C (113°F), and not all of Williams' (1956) cultures fermented both inulin and raffinose. Some strains of *Str. salivarius* do hydrolyse urea (Colman, 1970). Cultures of *Str. salivarius* do not form peroxide during growth (Carlsson, 1968).

*Str. salivarius* is present in large numbers in saliva and these organisms seem to colonize the mucous membranes of the mouth preferentially and can readily be recovered in large numbers from sites such as the tongue (Krasse, 1954) or the hard palate (Colman et al., 1975). In an experimental system, cells of *Str. salivarius*, but not of *Str. mutans*, adhered readily to epithelial cells (Gibbons et al., 1972).

A homogeneous species for *Str. salivarius* can be defined in terms of physiological properties but this species is serologically heterogeneous (Williams, 1956). Extracts of many strains yield precipitates with group K sera (Williams, 1956) and occasional strains may also yield precipitates with serum of a second Lancefield group. Sherman et al. (1943) used precipitin tests to distinguish at least two serological types. Nearly all of the strains that yield precipitates with group K sera also carry Sherman's type I antigen (Montague & Knox, 1968) and in addition cross-react (Mirick et al., 1944a) with antisera prepared against streptococcus MG (syn. *Str. milleri* serotype III). Strains of Sherman's serological type II do not cross-react with group K sera (Montague & Knox, 1968) or with sera against streptococcus MG.

## Streptococcus milleri

The name *Str. milleri* was proposed by Guthof (1956) for the non-haemolytic streptococci that he isolated from dental abscesses and other suppurative lesions around the mouth. Other names have been used for this species. Streptococcus MG (Mirick et al., 1944b) is properly classified here (Colman & Williams, 1967) but the name is not a binomial. In the eighth edition of *Bergey's Manual* (Buchanan & Gibbons, 1974) streptococcus MG is placed in *Str. anginosus* with the small-colony ('minute') haemolytic streptococci of Lancefield group F and group G. This classification has much to commend it but the name was introduced by Andrewes and Horder (1906) for the variety of *Str. pyogenes* that causes scarlet fever.

In the mouth *Str. milleri* appears to be particularly numerous in the gingival crevice (Mejàre & Edwardsson, 1975). It has been isolated from bacteraemias of dental origin (Kraus et al., 1953) but it is an infrequent cause of endocarditis (Parker & Ball, 1975). *Str. milleri* does cause abscesses in and about the gastro-intestinal tract, and apparently also causes meningitis and brain abscess (Parker & Ball, 1975). Mejàre and Edwardsson (1975) found that Carlsson's sulphonamide medium (1967a), originally described for the isolation of *Str. mutans*, is a useful selective medium for *Str. milleri*. The resistance of streptococcus MG to sulphonamide was used by Thomas and his colleagues (1945) as a means of selective isolation. The physiological properties of *Str. milleri* are distinctive (Colman & Williams, 1972, 1973). Lactose, sucrose, trehalose and salicin are fermented. Aesculin and arginine are hydrolysed. Nearly all strains form acetoin from glucose and are resistant to the concentration of bacitracin used in identification discs. Most grow on 10 per cent bile agar. Extra-cellular polysaccharides are not produced from sucrose, hydrogen peroxide is not formed during growth in culture, growth does not occur at 45°C (113°F) or in 4 per cent sodium chloride broth, neither mannitol nor sorbitol are fermented and starch is not hydrolysed. On horse blood agar colonies are non-haemolytic or are surrounded by narrow zones of greening.

Again in *Str. milleri* a physiologically homogeneous species is serologically heterogeneous. Cross-reactions are found with sera of the Lancefield series. Reaction with an F serum is the most common but a precipitin line might be found with a serum of one of the groups A, C, G, or K (Ottens & Winkler, 1962; Perch & Lind, 1968). Lancefield and Hare (1935) based their serological group F on strains isolated by Long and Bliss (1934). Bliss (1937) showed that some strains of the small colony haemolytic streptococci belonged to Lancefield group F, others to group G. Type antibodies were readily produced and Bliss's type I antigen was present in representatives of group F and group G. Her serotype II was found among strains that carried the F antigen. Later the same type II was found among haemolytic strains that cross-reacted with group A sera (Jablon et al., 1965), as well as among strains lacking a recognized group antigen (Ottens & Winkler, 1962).

The Bliss type I antigen is also found among non-haemolytic strains that fall within the range of characteristics of *Str. milleri*, and these might also carry the F antigen or lack an established group antigen. The Bliss type II antigen does not seem to occur among non-haemolytic strains. The Bliss types I and II were adopted by Ottens and Winkler (1962) as their types I and II respectively. Ottens and Winkler's type III, which is not found among haemolytic strains, is the type antigen of streptococcus MG (Mirick et al., 1944a) which is in turn very similar to Sherman's type I antigen of *Str. salivarius* (Willers et al., 1964). The Ottens type III antigen is found in strains of *Str. milleri* that might react with F or C grouping sera, and is also found among strains of this species that lack an established group antigen. The Ottens types IV and V are found among strains of *Str. milleri* and are there often associated with the Lancefield F antigen. de Moor and Michel (1968) examined strains that did not react with sera of the Ottens and Winkler or Lancefield series and have also (Michel et al., 1967) introduced a numerical cipher for these antigenic carbohydrates.

There is much to be said for bringing the so-called 'minute' haemolytic streptococci into *Str. milleri*, perhaps as a distinct variety. There are common antigens and also genetic similarities (Colman, 1969). In physiological properties differences are however seen between the haemolytic and non-haemolytic strains. The growth of the haemolytic strains on complex media is often stimulated by carbon dioxide; these strains differ from the non-haemolytic strains in their ability to produce hyaluronidase, but may fail to hydrolyse aesculin or ferment lactose.

## Streptococcus mitior

The name *Str. mitior* is used here for a species with a more restricted definition (Williams, 1973) than is applied to *Str. mitis*. The concept of *Str. mitior* begins with Schottmüller (1903) who isolated greening streptococci from the mouth, pharynx and gastro-intestinal tract. His organism formed very long chains that flocculated to the bottom of broth cultures, they also clotted milk but grew poorly at 22°C (71·6°F). Andrewes and Horder (1906) described their *Str. mitis* as growing in short chains, failing to clot milk and flourishing in cultures maintained at 20°C (68°F). Wirth (1926) examined some of Schottmüller's strains and observed additional properties such as failure to grow in the presence of a bile salt and death of the culture when heated to 60°C (140°F).

In a study of cell wall composition (Colman & Williams, 1965) some strains stood apart because rhamnose, the sugar forming the skeleton of many Lancefield group antigens, was not present in the cell walls. These strains formed a homogeneous cluster in computer studies (Colman, 1968) and, within the limits of the reciprocal transformation test, there was genetic similarity between strains (Colman, 1969). The definition of *Str. mitior* was emended (Colman, 1970; Colman and Williams, 1972, 1973) to include greening and haemolytic strains that ferment lactose and sucrose, that might ferment trehalose or raffinose, that rarely ferment salicin and inulin, and which do not ferment mannitol or sorbitol. Hydrogen peroxide is released during growth but neither arginine nor aesculin are hydrolysed. Cultures do not grow at 45°C (113°F), survive 60°C (140°F) for 30 min or grow in the presence of either 4 per cent sodium chloride or 10 per cent bile. The cell walls of these strains lack rhamnose but ribitol teichoic acid is usually present in appreciable quantities. The process of identification of these strains does not however require the analysis of isolated cell walls (Cowan, 1974) and if required the presence or absence of rhamnose can be ascertained (Gibbons, 1955) in extracts made by Fuller's (1938) formamide method from whole cells.

Strains that fit this description of *Str. mitior* form Carlsson's (1968) group VA and Liljemark and Gibbon's (1972) *Str. miteor*. The haemolytic and greening strains designated serological group O by Boissard and Wormald (1950) come within this species as do some strains that cross-react with sera of group M or group K. As mentioned in the section dealing with *Str. sanguis* some dextran-forming strains are a useful variety of *Str. mitior*.

In contrast to other species, such as *Str. salivarius*, *Str. sanguis* and *Str. mutans* which are most numerous in restricted sites in the mouth, *Str. mitior* is present in large numbers in dental plaque and also on mucosal surfaces such as the tongue and palate (Liljemark & Gibbons, 1972; Colman et al., 1975). Representatives of this species formed 14 per cent of the strains in Parker and Ball's (1975) series of strains isolated from endocarditis, and therefore possess some capacity for harm.

## Streptococcus pneumoniae

Techniques for the isolation and identification of *Str. pneumoniae* have been established by critical practice over many years (Austrian, 1970). These techniques are of current clinical importance partly because of the recognition of penicillin-resistant strains (Hansman, et al., 1974). The pneumococcus can often be isolated, before antibiotic therapy, by blood culture of patients with pneumococcal pneumonia (Austrian & Gold, 1964). A rapid method for diagnosis of pneumococcal bacteraemia was described by Dorff et al. (1971) who demonstrated capsular material in serum by immunoelectrophoresis. Another technique that does not require culture is the capsular precipitin reaction on smears using polyvalent serum (Lund & Rasmussen, 1966) and this reaction is more reliable than a simple Gram stain (Merrill et al., 1973).

Isolation is conveniently achieved on blood agar, and if only one plate is used this should be incubated in a mixture of air and carbon dioxide (Austrian & Collins, 1966). Overgrowth of pneumococci on blood agar by Gram-negative bacteria can be prevented by addition of nalidixic acid and polymyxin to the medium (Ellner et al., 1966). Inoculation of white mice with sputum is a sensitive method for the isolation of most serotypes of pneumococci and is of particular value with specimens containing few of these bacteria.

The two most widely used diagnostic tests for the pneumococci are sensitivity to optochin and lysis by bile salts and these characteristics are seen together only in *Str. pneumoniae*. The zone of inhibition produced by optochin is somewhat reduced when the colonies are incubated in a mixture of air and carbon dioxide (Ragsdale & Sandford, 1971). When grown in air on blood agar pneumococci are surrounded by zones of greening and also liberate hydrogen peroxide into the agar (Austrian, 1970). When incubated anaerobically on blood agar colonies of pneumococci are surrounded by zones of complete lysis caused by activation of the pneumolysin. This lysin apparently shows cross-neutralization with streptolysin O in hyperimmune horse sera but the same reaction is not seen with human sera (Bernheimer, 1972) suggesting that a pneumococcal infection is perhaps not followed by an increased antistreptolysin O titre.

## *Strains of uncertain position*

There are more species among the viridans streptococci of man than those described above. Carlsson's (1968) group IV strains are perhaps one such taxonomic unit and others perhaps are to be found among the strains that stand apart in any collection of cultures.

## Test methods

The test methods suitable for the study of the viridans streptococci are available, except for five properties, in Cowan (1974). Differential tests for this group of organisms are shown in Table 6.1 (pp. 180–1).

Extracellular polysaccharide production from sucrose can be demonstrated in the medium of Bailey and Oxford (1958a) by methods described by Hehre and Neill (1946). Autoclave solutions containing 17 per cent sucrose and, for control tubes, 17 per cent glucose. Add 1 ml of 11 per cent potassium carbonate (sterilized by filtration) to each 100 ml of sugar solution. Add an equal volume of a sugar solution to the sterile base which contains 2·8 per cent tryptone, 1 per cent yeast extract and 4 per cent sodium acetate. Inoculate with a drop of overnight broth culture and after incubation for 3–5 days dilute the culture supernatant 1 : 10 in a 10 per cent sodium acetate solution. Add 1·2 and 2·5 volumes of ethanol to separate tubes containing diluted sucrose broth culture fluid and add 2·5 volumes of ethanol to a single tube of supernatant from the control culture containing glucose. Levan is precipitated by 2·5 volumes of ethanol but not by 1·2 volumes and dextran is precipitated by both 1·2 and 2·5 volumes. Alternatively a series of dilutions of culture fluid can be made in a neutral buffer and the presence of dextran can be demonstrated by a tube precipitin test with *Str. pneumoniae* type II antiserum.

Fermentation reactions can be carried out in phenol red peptone water supplemented with 0·3 per cent beef extract powder, 0·1 per cent agar and 0·1 per cent $K_2HPO_4$, and containing 1 per cent of the substrate.

Hydrogen peroxide production by growing cultures can be detected in a medium, slightly modified from that described by Whittenbury (1964). Horse blood is lysed by repeated freezing and thawing and 3 ml of the lysed blood is added to each 100 ml of a sterile and melted blood agar base at 45°C (113°F). Mix vigorously and steam for 15 min with occasional mixing. Cool again to 45°C (113°F) and add 5 ml of an O-dianisidine solution to each 100 ml of base. The O-dianisidine is prepared by steaming for 10 min 0·5 g of the salt in 100 ml of distilled water with 0·3 ml of 36 per cent hydrochloric acid. Colonies producing hydrogen peroxide are surrounded by blackened medium. The results of the test are in part dependent on the agar base and fewer positive reactions are observed with D.S.T. agar (Oxoid)—(B. Mejàre, *personal communication*, 1974).

Starch hydrolysis in broth was described by Raibaud and his colleagues (1961). The medium contains 1 per cent tryptone, 0·5 per cent yeast extract and 0·2 per cent soluble starch and cultures are incubated for 5 days after inoculation with an overnight broth culture. To each 3 ml of medium are added 0·2 ml of a 12 per cent ferric chloride solution (0·2 ml of 36 per cent HCl is added to each 100 ml of the ferric chloride solution) and then 0·2 ml of Lugol's iodine solution.

Urea hydrolysis can be detected (Raibaud et al., 1961) by growing the cells for 5 days in a medium containing 1 per cent tryptone, 0·5 per cent yeast extract, 0·1 per cent glucose and to which has been added a solution of urea, sterilized by filtration, to give a concentration of 0·6 per cent. Free ammonia is detected with Nessler's reagent after the culture fluid has been diluted 1 : 10 with distilled water. Residual urea is detected by adding 2 volumes of a 1 per cent solution of xanthydrol in glacial acetic acid to culture fluid and then heating for 1 min in a boiling water bath. Any residual urea is precipitated by the reagents.

## References

Abercrombie, G. F. & Scott, W. M. (1928) A case of infective endocarditis due to *Streptococcus mutans*. *Lancet*, **2**, 697.

Andrewes, F. W. (1930) Note on the fermentation of starch by certain haemolytic streptococci. *J. Path. Bact.*, **33**, 145.

Andrewes, F. W. & Horder, T. J. (1906) A study of the streptococci pathogenic for man. *Lancet*, **2**, 708, 775, 852.

Austrian, R. (1970) *In: Manual of Clinical Microbiology*. J. E. Blair, E. H. Lennette & J. P. Truant, Eds p. 69, Bethesda: American Society for Microbiology.

Austrian, R. & Collins, P. (1966) Importance of carbon dioxide in the isolation of pneumococci. *J. Bact.*, **92**, 1281.

Austrian, R. & Gold, J. (1964) Pneumococcal bacteremia with especial reference to bacteremic pneumococcal pneumonia. *Ann. intern. Med.* **60**, 759.

Bailey, R. W. & Oxford, A. E. (1958a) Pre-requisites for dextran production by *Streptococcus bovis*. *Nature*, Lond. **182**, 185.

Bailey, R. W. & Oxford, A. E. (1958b) A quantitative study of the production of dextran from sucrose by rumen strains of *Streptococcus bovis*. *J. gen. Microbiol.*, **19**, 130.

Bernheimer, A. W. (1972) *In: Streptococci and Streptococcal Diseases*, pp. 19, 52, L. W. Wannamaker & J. M. Matsen, Eds. New York: Academic Press.

Bleiweis, A. S., Craig, R. A., Zinner, D. D. & Jablon, J. M. (1971) Chemical composition of purified cell walls of cariogenic streptococci. *Infect. Immun.*, **3**, 189.

Bliss, E. A. (1937) Studies upon minute hemolytic streptococci. III. Serological differentiation. *J. Bact.*, **33**, 625.

Boissard, J. M. & Wormald, P. J. (1950) A new group of haemolytic streptococci for which the designation "Group O" is proposed. *J. Path. Bact.*, **62**, 37.

Bowden, G. H. & Hardie, J. M. (1972) Stable heat and acid antigens common to species of *Streptococcus sanguis*. *J. dent. Res.*, **51**, 1257.

Brown, J. H. (1919) *The Use of Blood Agar for the Study of Streptococci.* Monograph of the Rockefeller Institute of·Medical Research No. 9.

Buchanan, R. E. & Gibbons, N. E. (1974) Bergey's Manual of Determinative Bacteriology. 8th ed., pp. 500, 502. Baltimore: Williams & Wilkins Co.

Carlsson, J. (1967a) A medium for isolation of *Streptococcus mutans. Archs oral Biol.,* **12**, 1657.

Carlsson, J. (1967b) Presence of various types of non-haemolytic streptococci in dental plaque and in other sites of the oral cavity in man. *Odont. Revy,* **18**, 55.

Carlsson, J. (1968) A numerical taxonomic study of human oral streptococci. *Odont. Revy,* **19**, 137.

Clarke, J. K. (1924) On the bacterial factor in the aetiology of dental caries. *Br. J. exp., Path.,* **5**, 141.

Colman, G. (1967) Aerococcus-like organisms isolated from human infections. *J. clin. Path.,* **20**, 294.

Colman, G. (1968) The application of computers to the classification of streptococci. *J. gen. Microbiol.,* **50**, 149.

Colman, G. (1969) Transformation of viridans-like streptococci. *J. gen. Microbiol.,* **57**, 247.

Colman, G. (1970) *The Classification of Streptococcal Strains.* Ph.D. thesis: University of London.

Colman, G., Beighton, D., Chalk, A. J. & Wake, S. (1975) Cigarette smoking and the microbial flora of the mouth. *Aust. dent. J.,* (in press).

Colman, G., Wake, S., Chalk, A. J. & Beighton, D. (1974) Selective media for plaque bacteria. *J. dent. Res.,* **53**, 710.

Colman, G. and Williams, R. E. O. (1965) The cell walls of streptococci. *J. gen. Microbiol.,* **41**, 375.

Colman, G. & Williams, R. E. O. (1967) Classification of non-haemolytic streptococci. *Int. J. syst. Bact.,* **17**, 306.

Colman, G. & Williams, R. E. O. (1972) *In: Streptococci and Streptococcal Diseases.* p. 281. L. W. Wannamaker & J. M. Matsen, Eds, New York: Academic Press.

Colman, G. & Williams, R. E. O. (1973) *In: Recent Advances in Clinical Pathology Series,* 6, p. 293. S. C. Dyke, Ed. Edinburgh: Churchill Livingstone.

Cowan, S. T. (1947) *Cowan and Steel's Manual for the Identification of Medical Bacteria.* pp. 51–53. London: Cambridge University Press.

Crowe, H. W. (1927) The differentiation and classification of the non-haemolytic streptococci by the use of Crowe's medium. *Ann. Pickett–Thomson Res. Lab.* 3, 251.

Cybulska, J. & Pakula, R. (1963) Streptococcal polyglucosidases. II. The effect of some factors on enzymatic synthesis of dextran from sucrose in cell free media. *Exp. Med. Microbiol.,* **15**, 285.

Dorff, G. J., Coonrod, J. D. & Rytel, M. W. (1971) Detection by immunoelectrophoresis of antigen in sera of patients with pneumococcal bacteraemia. *Lancet,* **1**, 578.

Dunican, L. K. & Seeley, H. W. (1962) Starch hydrolysis by *Streptococcus equinus. J. Bact.,* **83**, 264.

Ebisu, S., Misaki, A., Kato, K. & Kotani, S. (1974) The structure of water-insoluble glucans of cariogenic *Streptococcus mutans,* formed in the absence and presence of dextranase. *Carbohydrate Res.,* **38**, 374.

Eisenberg, R. J., Elchisak, M. & Lai, C. (1974) Glycogen accumulation by pleomorphic cells of *Streptococcus sanguis*. *Biochem. biophys. Res. Commun.,* **57**, 959.

Ellner, P. D., Stoessel, C. J., Drakeford, E. & Vasi, F. (1966) A new culture medium for medical bacteriology. *Am. J. clin. Path.,* **45**, 502.

Facklam, R. R. (1972) Recognition of group D streptococcal species of human origin by biochemical and physiological tests. *Appl. Microbiol.,* **23**, 1131.

Facklam, R. R. (1974) Characteristics of *Streptococcus mutans* isolated from human dental plaque and blood. *Int. J. syst. Bact.,* **24**, 313.

Fuller, A. T. (1938) The formamide method for the extraction of polysaccharides from haemolytic streptococci. *Br. J. exp. Path.,* **19**, 130.

Gibbons, M. N. (1955) The determination of methylpentoses. *Analyst, Lond.,* **80**, 268.

Gibbons, R. J. (1972) *In: Streptococci and Streptococcal Diseases* p. 371. L. W. Wannamaker & J. M. Matsen, Eds. New York: Academic Press.

Gibbons, R. J., van Houte, J. & Liljemark, W. F. (1972) Parameters that effect the adherence of *Streptococcus salivarius* to oral epithelial surfaces. *J. dent. Res.,* **51**, 424.

Gledhill, W. E. & Casida, L. E. Jr. (1969) Predominant catalase-negative soil bacteria. I. Streptococcal population indigenous to soil. *Appl. Microbiol.,* **17**, 208.

Gold, O. G., Jordan, H. V. & van Houte, J. (1973) A selective medium for *Streptococcus mutans. Archs oral. Biol.,* **18**, 1357.

Guggenheim, B. (1975) Isolation and properties of a dextranase isolated from *Streptococcus mutans* OMZ 176. *Caries Res.,* **9**, 309.

Guthof, O. (1956) Ueber pathogene 'vergrünende Streptokokken'. Streptokokken-Befunde bei dentogenen Abszessen und Infiltraten im Bereich der Mundhohle. *Zentbl. Bakt., ParasitKde,* (Abt. 1), **166**, 553.

Hansman, D., Devitt, L., Miles, H. & Riley, I. (1974) Pneumococci relatively insensitive to penicillin in Australia and New Guinea. *Med. J. Aust.,* **2**, 353.

Hare, R. (1935) *Studies on the Classification of Haemolytic Streptococci, the Aetiology of Infections, and the Immunity of Human Beings for these Organisms.* M.D. thesis: University of London.

Hehre, E. J. & Neill, J. M. (1946) Formation of serologically reactive dextrans by streptococci from subacute bacterial endocarditis. *J. exp. Med.,* **83**, 147.

Heim, L. (1924) Milchsaure und andere Streptokokken. *Z. Hyg. InfektKrankh.,* **101**, 104.

van Houte, J., Jordan, H. V. & Bellack, S. (1971) Proportions of *Streptococcus sanguis,* an organism associated with subacute bacterial endocarditis, in human feces and dental plaque. *Infect. Immun.,* **4**, 658.

Iverson, W. G. & Millis, N. F. (1974) A method for the detection of starch hydrolysis by bacteria. *J. appl. Bact.,* **37**, 443.

Jablon, J. M., Brust, B. & Saslaw, M. S. (1965) β-haemolytic streptococci with group A and type II carbohydrate antigens. *J. Bact.,* **89**, 529.

Keyes, P. H. (1960) The infectious and transmissible nature of experimental dental caries. Findings and implications. *Archs oral. Biol.,* **1**, 304.

Kiel, P. & Skadhauge, K. (1973) Studies on mannitol-fermenting strains of *Streptococcus bovis. Acta Path. Microbiol. Scand.,* B. **81**, 10.

Krasse, B. (1954) The proportional distribution of *Streptococcus salivarius* and other streptococci in various parts of the mouth. *Odont. Revy,* **5**, 203.

Kraus, F. W., Casey, D. W. & Johnson, V. (1953) The classification of non-hemolytic streptococci recovered from bacteremia of dental origin, *J. dent. Res.*, **32**, 613.

Lancefield, R. C. & Hare, R. (1935) The serological differentiation of pathogenic and non-pathogenic strains of hemolytic streptococci from parturient women. *J. exp. Med.*, **61**, 335.

Liljemark, W. F. & Gibbons, R. J. (1972) Proportional distribution and relative adherence of *Streptococcus miteor* (*mitis*) on various surfaces in the human oral cavity. *Infect. & Immun.*, **6**, 852.

Llory, H., Guillo, B. & Frank, R. M. (1971) A cariogenic *Actinomyces viscous* —a bacteriological and gnotobiotic study. *Helv. odont. Acta*, **15**, 134.

Long, P. H. & Bliss, E. A. (1934) Studies upon minute hemolytic streptococci. I. The isolation and cultural characteristics of minute beta hemolytic streptococci. *J. exp. Med.*, **60**, 619.

Lund, E. & Rasmussen, P. (1966) Omni-serum: a diagnostic pneumococcus serum reacting with the 82 known types of pneumococcus. *Acta. Path. Microbiol. Scand.*, **68**, 45.

Maclean, I. H. (1927) The bacteriology of dental caries. *Proc. R. Soc. Med.*, **20**, 873.

Medrek, T. F. & Barnes, E. M. (1962) The influence of the growth medium on the demonstration of a group D antigen in faecal streptococci. *J. gen. Microbiol.*, **28**, 701.

Mejàre, B. & Edwardsson, S. (1975) Observations on the characteristics of *Streptococcus milleri* and its occurrence as a regular inhabitant of the human mouth. *Caries Res.*, **9**, 308.

Merrill, C. W., Gwaltney, J. M., Hendley, J. O. & Sande, M. A. (1973) Rapid identification of pneumococci. Gram stain vs the Quellung reaction. *New Engl. J. Med.*, **288**, 510.

Michel, M. F., de Moor, C. E., Ottens, H., Willers, J. M. N. & Winkler, K. C. (1967) A note on the nomenclature of certain polysaccharides resembling the group antigens of streptococci. *J. gen. Microbiol.*, **49**, 49.

Mieth, H. (1962) Untersuchungen über das Vorkommen von Enterokokken bei Tieren und Menschen. IV. Mitteilung Die Streptokokken flora in den Faeces von Pferden. *Zentbl. Bakt., ParasitKde*, (Abt. 1). **185**, 166.

Mirick, G. S., Thomas, L., Curnen, E. C. & Horsfall, F. L. (1944a) Studies on a non-hemolytic streptococcus isolated from the respiratory tract of human beings. III. Immunological relationship of Streptococcus MG to *Streptococcus salivarius* type 1. *J. exp. Med.*, **80**, 431.

Mirick, G. S., Thomas, L., Curnen, E. C. & Horsfall, F. L. (1944b) Studies on a non-hemolytic streptococcus isolated from the respiratory tract of human beings. I. Biological characteristics of Streptococcus MG. *J. exp. Med.*, **80**, 391.

Montague, E. A. & Knox, K. M. (1968) Antigenic components of the cell wall of *Streptococcus salivarius*. *J. gen. Microbiol.*, **54**, 237.

de Moor, C. E., van Houte, J. & de Stoppelaar, J. D. (1968) Endocarditis lenta en caries streptokokken. *Streptococcus mutans. Versl. Meded. Volksgezondh.*, p. 325

de Moor, C. E. & Michel, M. F. (1968) Groep F. streptokokken. Biochemische reactics, serologie en immunochemie. *Versl. Meded. Volksgezondh.*, p. 308.

Munch-Petersen, E. (1954) Note on *Streptococcus agalactiae. Int. Bull. bact. Nomencl. Taxon.,* **4**, 129.

Niven, C. F. Jr., Kiziuta, Z. & White, J. C. (1946) Synthesis of a polysaccharide from sucrose by Streptococcus s.b.e. *J. Bact.,* **51**, 711.

Niven, C. F. Jr., Smiley, K. L. & Sherman, J. M. (1941a) The production of large amounts of a polysaccharide by *Streptoccocus salivarius. J. Bact.,* **41**, 479.

Niven, C. F. Jr., Smiley, K. L. & Sherman, J. M. (1941b) The polysaccharides synthesized by *Streptococcus salivarius* and *Streptococcus bovis. J. biol. Chem.,* **140**, 105.

Niven, C. F. Jr., Washburn, M. R. & White, J. C. (1948) Nutrition of *Streptococcus bovis. J. Bact.,* **55**, 601.

Orla-Jensen, S. (1919). *The Lactic Acid Bacteria.* Copenhagen.

Orland, F. J., Blayney, J. R., Harrison, R. W., Reyniers, J. A., Trexler, P. C., Ervin, R. F., Gordon, H. A. & Wagner, M. (1955) Experimental caries in germ free rats inoculated with enterococci. *J. Am. dent. Ass.,* **50**, 259.

Ottens, H. & Winkler, K. C. (1962) Indifferent and haemolytic streptococci possessing group-antigen F. *J. gen. Microbiol.,* **28**, 181.

Parker, M. T. & Ball, L. C. (1975) Streptococci associated with systemic disease in man. Pathological Society of Great Britain and Ireland, 113th meeting. Pre-printed abstracts.

Perch, B., Kjems, E. & Ravn, T. (1974) Biochemical and serological properties of *Streptococcus mutans* from various human and animal sources. *Acta Path. Microbiol. Scand. Sect.* B. **82**, 357.

Perch, B. & Lind, K. (1968) Immunological relationships of Streptococcus MG to group K streptococcus, strain K 4a. *Acta Path. Microbiol. Scand.,* **73**, 220.

Porterfield, J. S. (1950) Classification of the streptococci of subacute bacterial endocarditis. *J. gen. Microbiol.,* **4**, 92.

Ragsdale, A. R. & Sandford, J. P. (1971) Interfering effect of incubation in carbon dioxide on the identification of pneumococci by optochin discs. *Appl. Microbiol.,* **22**, 854.

Raibaud, P., Caulet, M., Galpin, J. V. & Mocquot, G. (1961) Studies on the bacterial flora of the alimentary tract of pigs. II. Streptococci: selective enumeration and differentiation of the dominant group. *J. appl. Bact.,* **24**, 285.

Ravreby, W. D., Bottone, E. J. & Keusch, G. T. (1973) Group D streptococcal bacteremia, with emphasis on the incidence and presentation of infections due to *Streptococcus bovis. New Engl. J. Med.,* **289**, 1400.

Rosan, B. (1973) Antigens of *Streptococcus sanguis. Infect. Immun.,* **7**, 205.

Rosan, B. & Eisenberg, R. J. (1973) Morphological changes in *Streptococcus sanguis* associated with growth in the presence of oxygen. *Arch. oral Biol.,* **18**, 1441.

Safford, C. E., Sherman, J. M. & Hodge, H. M. (1937) *Streptococcus salivarius. J. Bact.,* **33**, 263.

Schottmüller, H. (1903) Die Artunterscheidung der für den Menschen pathogenen Streptokokken durch Blutagar. *Münch. med. Wschr.,* **50**, 849, 909.

Seeley, H. W. & Dain, J. A. (1960) Starch hydrolyzing streptococci. *J. Bact.,* **79**, 230.

Shattock, P. M. F. (1949) The streptococci of group D; the serological grouping of *Streptococcus bovis* and observations on serologically refractory group D strains. *J. gen. Microbiol.,* **3**, 80.

Sherman, J. M. (1937) The streptococci. _Bact. Rev._, **1**, 3.

Sherman, J. M., Niven, C. F. Jr. & Smiley, K. L. (1943) _Streptococcus salivarius_ and other non-hemolytic streptococci of the human throat. _J. Bact._, **45**, 249.

Sherman, J. M. & Stark, P. (1931) Streptococci which grow at high temperatures. _J. Bact._, **22**, 275.

Sims, W. (1964) A simple test for differentiating _Streptococcus bovis_ from other streptococci. _J. appl., Bact._, **27**, 432.

Smith, D. G. & Shattock, P. M. F. (1962) The serological grouping of _Streptococcus equinus. J. gen. Microbiol._ **29**, 731.

Staat, R. H. & Schachtele, C. F. (1974) Evaluation of dextranase production by the cariogenic bacterium _Streptococcus mutans. Infect. Immun.,_ **9**, 467.

de Stoppelaar, J. D., van Houte, J. & de Moor, C. E. (1967) The presence of dextran-forming bacteria, resembling _Streptococcus bovis_ and _Streptococcus sanguis,_ in human dental plaque. _Archs. oral Biol.,_ **12**, 1199.

de Stoppelaar, J. D., van Houte, J. & Backer Dirks, O. (1969). The relationship between extracellular polysaccharide-producing streptococci and smooth surface caries in 13-year old children. _Caries Res.,_ **3**, 190.

Swan, A. (1954) The use of a bile-aesulin medium and of Maxted's technique of Lancefield grouping in the identification of enterococci (group D streptococci). _J. clin. Path.,_ **7**, 160.

Switzer, R. E. & Evans, J. B. (1974) Evaluation of selective media for enumeration of group D streptococci in bovine feces. _Appl. Microbiol.,_ **28**, 1086.

Thomas, L., Mirick, G. S., Curnen, E. C., Ziegler, J. E. Jr. & Horsfall, F. L. Jr. (1945) Studies on primary atypical pneumonia; II Observations concerning relationship of non-hemolytic streptococcus to disease. _J. clin. Invest.,_ **24**, 227.

Thomson, D. & Thomson, R. (1929) Researches on the role of streptococci in oral and dental sepsis. _Ann. Pickett-Thomson Res. Lab.,_ **5**, 1.

Washburn, M. R., White, J. C. & Niven, C. F. Jr. (1946) Streptococcus s.b.e.: immunological characteristics. _J. Bact.,_ **51**, 723.

Whittenbury, R. (1964) Hydrogen peroxide formation and catalase activity in the lactic acid bacteria. _J. gen. Microbiol.,_ **35**, 13.

Willers, J. M. N., Ottens, H. & Michel, M. F. (1964) Immunological relationship between streptococcus, MG, F111 and _Streptococcus salivarius J. Gen. Microbiol.,_ **37**, 425.

Williams, R. E. O. (1956) _Streptococcus salivarius_ (vel. _hominis_) and its relation to Lancefield's group K. _J. Path. Bact.,_ **72**, 15.

Williams, R. E. O. (1973) Benefit and mischief from commensal bacteria. _J. Clin. Path.,_ **26**, 811.

Wirth, E. (1926) Zur Kenntnis der Streptokokken. _Zentbl. Bakt., ParasitKde,_ (Abt. 1), **99**, 266.

# 7 | Microbiology of Medicinal Products

G. SYKES

In recent years interest has been stimulated in the microbiological quality of pharmaceutical products, largely as a result of the publication of a report by the Swedish Health Laboratory (Kallings et al., 1965; Kallings et al., 1966) and in this country by bringing into effect the Guide to Good Pharmaceutical Manufacturing Practice (GMP) (1971) made under the Medicines Act, 1968. The former was concerned with the occurrence of pathogenic bacteria, notably *Salmonella muenchen* and *Salmonella bareilly*, in thyroid tablets, and suggested limits for the bacterial content of all drugs. The Medicines Act (1968) is concerned with the overall quality and safety of all products used for medicinal purposes and it gives comprehensive definitions of 'medicinal products' and 'medicinal purposes'. Besides these publications numerous other reports have appeared on the occurrence and consequences of undesirable microbial contaminants in a variety of pharmaceutical formulations, including creams, emulsions, antiseptic solutions, antibiotics, eye preparations and infusion fluids, as well as in a range of powders and tablets. Perhaps the most important of these reports in the present context are those describing several cases involving loss of sight due to the use of eye drops contaminated with *Pseudomonas aeruginosa* (Ayliffe et al., 1966) and the Clothier Report (*Command*, 1972) which investigated the death of several patients following the administration of contaminated infusion fluids. These reports serve to underline the significance of microbial contamination in medicinal products, and its possibly tragic consequences, and to emphasize the need for strict control of such products.

It is interesting that the Medicines Act (1968) and the Regulations arising from it do not mention microbial contamination, except obliquely in its references to sterile products, nevertheless the need for the

*199*

H

control of such contamination is strongly implied by the emphasis placed on purity and quality in relation to medicinal products.

## The Medicines Act (1968)

One of the main purposes of the Medicines Act is to control by licence the manufacture and distribution of all types of medicinal products, always with the primary object in mind of the greater safety of the consumers. The Act covers medicines administered to humans and animals, the standards for which are the same. In this report discussion however will be centred on medicines for human consumption.

Until the advent of the Medicines Act it was possible for anyone to manufacture almost any medicine under any conditions, the only restrictions being those imposed by the Therapeutic Substances Act (1956), which controlled the production of medicines which cannot be controlled by chemical or physical means, and the Pharmacy and Poisons Act (1933). There was, therefore, very little control over the efficacy, purity or consistency of medicinal products and it does not require much imagination to see that this lack of control could easily hazard the safety of the consumer or patient. The Act was designed to eliminate these hazards and variables and so ensure that all drugs and their formulations, together with certain other materials and appliances defined as medicines, were properly standardized and controlled and so took on more of the nature of 'known quantities'.

The first steps in this direction had been taken before 1968 when the Committee on Safety of Drugs, the Dunlop Committee, was set up. This body was established to control, on a voluntary basis, the production of new drugs. From this beginning was developed, under the Act, the Medicines Commission which was appointed to advise Ministers on all aspects of the application of the Act, and its various committees and sub-committees, one of which is the Committee on Safety of Medicines. This approach by no means usurps the authority of the British Pharmacopoeia (BP) and other official compendia in establishing standards for drugs and their formulation, neither does it restrict the freedom of the medical practitioner to prescribe drugs of his choice; it seeks rather to reinforce them and so ensure that the drug of choice is of the correct and approved quality.

The Medicines Act is administered through a series of Regulations, or Statutory Instruments, each of which must be approved by Parliament before it can become effective. It differentiates between 'manufacture' and 'dispensing' as applied to medicines. The latter in its simplest terms, can be described as the preparation of a medicine against a prescription by a medical practitioner for administration to a named patient, and it must be carried out by or under the supervision of a pharmacist or by a medical practitioner. The Act, therefore, underlines the professional and fundamental duty of the pharmacist to use his skill in the dispensing of medicines.

Although the Act is concerned with the manufacture and distribution of medicines by the pharmaceutical industry it is recognized that a large proportion of medicines used today are made in hospital pharmacies and that many of them are made for stock rather than as individual dispensed items. Thus they can be considered as manufactured items, and there is no reason why they should not be controlled to the same standards as those which obtain in industry and the departments manufacturing them licensed or approved in the same way as commercial establishments.

The control of manufacture is exerted through two types of licence— the product licence and the manufacturing licence. The product licence is an official authorization to promote for sale a specific medicinal product. The licence is only granted when the licensing authority is satisfied that: (a) the ingredients are of an acceptable quality; (b) the formulation yields a stable product of the right activity or potency and is free from undue toxicity; (c) the checks during processing and the final testing are adequate to ensure the required purity of the product; (d) the presentation, including labelling, conform to the required standards.

At present only a few products on the market have full product licences as described above. The large majority were on sale before the implementation date (September 1971) and so were issued with a product licence of right. The requirements for this type of licence are far less exacting than those for a full product licence but it is the intention to review all products on the market and eventually to require all pro-products to conform to the higher standard.

The manufacturing licence is an official authorization for a manufacturer to make for sale products which already have product licences or product licences of right. It is granted by the licensing authority on the basis of an assessment made by the Medicines Inspectorate of the Department of Health and Social Security of the manufacturer's capability to make his chosen range of products. Contrary to the situation in the past, inspection now covers hospital as well as industrial manufacturing units. The assessment is based on the requirements of Section 9 (5) of the Act as set out in the GMP Guide (1971) and is considered under the main headings of: (1) premises; (2) equipment; (3)

personnel; (4) cleanliness and hygiene; (5) adequate production procedures, documentation and records for the types of products made; and (6) quality control. Considerable emphasis is placed on quality control and on documentation and records.

The guide contains an Appendix for additional requirements for the manufacture of sterile medicinal products, a revision of which was published recently (1974). This revision does not alter the principles laid down in the original Guide; it is more expansive and explanatory, but still it does not quantify any of the recommendations. In fact, the intention of the Guide as a whole is to provide the manufacturer with a series of headings for him to consider with a view to establishing minimum acceptable standards. It should be readily understood that some specifications may be quite acceptable under certain conditions but inadequate under others.

Other countries, notably Australia, Canada, Sweden and the USA have also produced Guides or Codes in their own style, and so have the EFTA countries and WHO. The principles followed in each of the Guides are virtually the same, although the presentation is naturally different. Each puts appropriate emphasis on quality control and documentation and each has special sections dealing with sterile products.

## Non-Sterile Pharmaceutical Products

Microbial contaminants in non-sterile pharmaceutical products can originate from (Fig. 7.1):

(1). The raw or starting materials (including water).
(2). The equipment and containers used during manufacture.
(3). The air and general environment.
(4). The operators handling the materials.
(5). The containers and closures for the finished products.
(6). The users.

Of these, the most difficult to control microbiologically are the raw materials: it is also impossible for the manufacturer to control the user but he can protect his product to some extent by incorporating antimicrobial preservatives in the formulation. In the following paragraphs it is possible to give little more than a catalogue of the levels and types of contamination found under these headings.

**Fig. 7.1.** *Sources of microbial contamination.*

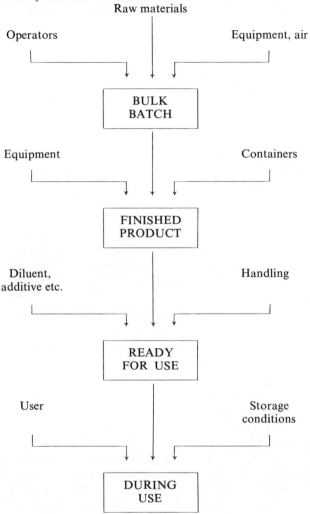

## Raw Materials

For the present study raw, or starting, materials can be grouped under three main headings :

(1). The drugs themselves, which may be further subdivided into :

(a) synthetic or semi-synthetic drugs and

(b) drugs of plant or animal origin.

(2). Diluents, excipients, lubricants and other additives.

(3). The water supply.

*Drugs.* Synthetic drugs in a dry state and kept under clean conditions can normally be expected to have low microbial counts, often less than 10/g. The organisms present are mainly bacterial and mould spores, with occasional cocci: Gram-negative, waterborne bacteria also occur and rarely obligate anaerobic bacilli. Naturally, there are exceptions to this general rule and calcium lactate and hydrocortisone are amongst the substances reported to contain up to several thousand viable bacteria per gram (Pedersen & Ulrich, 1968; Sykes, 1971). Even some of the non-sterile antibiotics, mainly antifungal agents, can contain many thousands of bacteria per gram, and their preparations in water can sometimes support growth.

Drugs of natural origin, both plant and animal, can also carry large numbers of bacteria, often in the range $10^5$–$10^7$/g (*Report*, 1971) and sometimes they include *E. coli* and salmonellae. Examples quoted in the literature (Sykes, 1971) include thyroid powder, digitalis leaf, senna fruit and carmine.

The processes of extraction and purification, involving, as they may well do, heat, low pH values and the use of alcohol or other organic solvents, reduce substantially the numbers of contaminating organisms in such drugs but recontamination can occur during subsequent stages of handling. Often this is from the crude drug, and results from the processing conditions being inadequately controlled. Animal products tend to be more susceptible to contamination than those from plants, and animal products are much more likely to carry organisms pathogenic to man, especially the salmonellae.

The natural earths, bentonite, kaolin and talc, can also carry large numbers of bacteria, depending on their origin and treatment, but most are harmless saprophytes. Nevertheless, clostridia are found, so that it is undesirable for such materials to be used in baby powders or in preparations, including cosmetics, which might be applied to the broken skin, unless they have been treated to eliminate these organisms. There is on record at least one case of neonatal death from tetanus caused by using a contaminated dusting powder.

*Excipients and other additives.* The many additives used in pharmaceutical formulations are also carriers of the same types of contaminants and, being the major constituents in many formulations, are significant in contributing to the microbiological quality of the finished product. They thus merit at least as much consideration as do the drugs. Starch is one of the commonest excipients used in tablet making, and Kallings and his co-workers (1965, 1966) focused attention on this when they found coliforms, staphylococci and other organisms in large numbers in tablets of barbiturates, tranquillizers, digitalis and other alkaloids.

Other workers broadly confirmed these findings and, perhaps not surprisingly, showed different rates of contamination between starches of different grade and origin. It is clear from more recent experimental evidence that the starch manufacturers have considerably improved their production methods and now there are usually no more than a few hundred viable organisms per gram, most of these are spores of *Bacillus* spp.

Other excipients such as lactose, glucose and sucrose carry quite low numbers of bacteria, generally less than 100/g, and it is rare for samples to contain more than 1 000/g. Substances used as lubricants in tablet manufacture, e.g. stearic acid and magnesium stearate, are similarly lightly contaminated.

Pharmaceutical formulations other than tablets can contain a variety of additives, including agar, gelatin, gum arabic and tragacanth, and again, because of the nature of their origin and the ways in which they might be processed and handled, the level of microbial contamination varies considerably. Gelatin, being used extensively in the confectionary trade, is usually well controlled during processing and so is relatively lightly contaminated; tragacanth on the other hand frequently carries salmonellae. The basic constituents of ointments and creams, such as cetyl alcohol, glycerol, lanolin, paraffins and oils, all have low viable counts often of less than 10/g.

*Water.* Water is not only the commonest ingredient in pharmaceutical products but it also presents some of the greatest microbiological problems. Products containing water are always more susceptible to microbial growth than are dry products and the growth is selective, favouring those organisms with less exacting nutritional requirements. Thus the waterborne pseudomonads and allied genera proliferate most readily but this does not preclude the possible growth of coliforms, salmonellae or *Ps. aeruginosa* as well as other potentially harmful organisms. In effect water plays a dual role: (a) it is the source of many of the contaminating bacteria and (b) it provides the menstruum for microbial growth.

Although it is accepted that, both microbiologically and chemically, tap water is unsuitable for use in pharmaceutical manufacture, the hazards associated with distilled water and purified (deionized) water can be even greater. Distilled water, immediately after condensation, is sterile and with proper aseptic collection will remain so. Generally it is collected simply under clean conditions, so that inevitably a few bacteria from the air and from the containers are present. Even in the apparent absence of nutrient materials growth may still occur, so that it is not surprising to find distilled water supplies, even after only a few days storage, with viable bacterial counts sometimes greater than $10^6$/ml. In addition the water may pass through plastic or metal pipe-lines to the receiving tank, and these lines, unless regularly cleaned and sterilized, can act as constant reservoirs of infection to the water passing through them. An example of this was the discovery, in a hospital, that water 'fresh from the still' and having passed through a short length of plastic tubing regularly had viable counts of the order of $10^5$ Gram-negative bacilli per ml, but when the plastic tubing was replaced every day with freshly sterilized material the level of con-tamination was reduced to less than 1 organism/ml. Membrane filtra-tion has also been recommended for removing bacterial cells from water and it does this effectively, but it does not remove the pyrogens which might also be present—neither does heat remove them. More-over, unless the filtration is undertaken as a sterile operation, the few remaining bacteria in the system and containers begin to grow, as already indicated.

In this connection it is also well to remember that the waterborne bacteria in general grow much better at 25°C (77°F) than at 37°C (98·6F). Thus the viable count at 25°C can be much higher than that at 37°C, sometimes by a factor of $10^3$.

The problem with deionized water is even greater because the ion exchange column acts as a reservoir for organisms which it absorbs from the original water supply. These organisms continue to grow and accumulate on the column, so that the issuing water becomes pro-gressively less and less satisfactory, with bacterial counts perhaps of the order of $10^3$ to $10^4$/ml. Mixed-bed columns are said to be even more susceptible in this respect than twin-bed columns. It is quite likely, therefore, that deionized water, even straight from a column, is in a worse state microbiologically than was the original supply, and this certainly is the case if the columns are used only intermittently. Even an overnight break, and more so a weekend one, could give rise to a substantial increase in the viable count. For these reasons deionized water is not suitable for use in the manufacture of pharmaceutical products unless it has been treated immediately before use to eliminate the contaminants. The British Pharmaceutical Codex (BPC) prescribes boiling and cooling immediately before use in the preparation of creams, but filtration or passage through a u.v. water treatment unit is equally satisfactory. A further problem with deionized water is the potential risk of the products becoming contaminated by resin particles.

*Other sources of contamination.* Among the other potential sources of contamination are the equipment and containers used during manufacture. Particular attention needs to be paid to such items as pipelines, valves and faucets, besides the bulk containers themselves and their covers. Unless they are kept scrupulously clean and protected from recontamination from the environment or from other materials they can act as prolific sources of micro-organisms for contaminating any preparation coming into direct contact with them. The variety of such contaminants can vary considerably and will depend largely on the nature of the product being processed through the plant. The main offenders are usually again the waterborne non-pigmented pseudomonads, but yeasts, moulds, bacterial spores and cocci, including *Staphylococcus aureus*, can also occur. Algae are, of course, always present but only occasionally responsible for heavy contamination. The proper procedure is to clean the whole plant immediately after use and flush with steam or hot water, repeating the procedure immediately before the next use. Unless such precautions are taken colonization can take place in the less accessible areas of the equipment which then act as a continuous source of infection. 'Dead ends', even if only a few inches long, valves and taps are particularly susceptible areas and plastic surfaces are worse offenders than metal ones, provided the latter are smooth and not scratched or corroded.

The numbers and types of airborne micro-organisms vary considerably with the environment and the activity taking place. Those most commonly occurring are again bacterial and mould spores, yeasts and cocci; Gram-negative bacteria are much less common. The same types of organisms are also found in the final containers and closures for the finished product but the numbers from these sources are not usually high. Glass containers are no better than plastic ones, and bottle washing, unless carefully controlled, can leave a container in a worse state microbiologically than before it was washed. Probably the best way of removing particles from the inside of a bottle is by blowing with filtered air.

The personnel themselves are always potential sources of contamination not only with commensals but also with *Staph. aureus* and streptococci as well as salmonellae, *E. coli* and other intestinal bacteria, and *Ps. aeruginosa*, the last of which is said to be carried by up to 12 per cent of normally healthy people. The risk here is two-fold; firstly from the operator during manufacture and, secondly, from the consumer during use: both have to be guarded against.

## Finished Products

A few preparations, for example, those containing alcohol, are self-disinfecting by the nature of their formulations, so that any microbial contamination introduced with the raw or starting materials or during manufacture is of no significance because it is rapidly killed. There are, however, many preparations in which microbial growth can take place rapidly and it is here that special precautions have to be taken. The greatest difficulties arise with aqueous preparations, including creams and emulsions, and often the problem can only be overcome by using a suitable antimicrobial preservative. Such a preservative should never be used to mask poor quality starting materials or bad manufacturing practices.

*Tablets, capsules and powders.* With the exception of drugs of natural or biological origin, the microbial contents of this group are generally not high but they vary in both numbers and types according to the nature of the drug and of the excipient used. Tablets of synthetic drugs usually have viable bacterial counts of less than 100 per tablet and often less than 10 per tablet. Most of the bacteria are aerobic spore formers of the *Bacillus subtilis* group and moulds, with occasional cocci. Reference has already been made to the relatively high counts found in drugs of natural origin and this is reflected in the counts found in tablets made with these drugs, examples of which are cited in Table 7.1.

If excessive moisture is present it is possible for growth of both bacteria and moulds to occur but bacteria cease to grow when the relative humidity (RH) falls below about 85 per cent; moulds cease to grow below about 75 per cent RH. It is possible, therefore, for tablets in a moist atmosphere to show discolouration and other spoilage due to mould growth. This can even occur in jars of tablets stored in such a way that one side of the jar is warmer than the other, thus causing moisture to condense on the cooler side. Dried products such as tablets, powders and capsules also have their problems because, although they do not allow the actual growth of bacteria, they can maintain them in a viable state for long periods so that, given suitable conditions, growth and spoilage can take place at almost any time in the future.

Mention should also be made of french chalks used in baby and

**Table 7.1.** *Bacterial content of some tablets*

| Tablet | Count/ tablet | Coliforms/ tablet | Comment |
|---|---|---|---|
| Digitalis | $10^2$–$10^5$ | 1–100 (*E. coli* in 1 sample) | Gram-negative rods up to $10^4$/tablet |
| Ergot | $10^2$–$10^3$ | 0:1 | |
| Thyroid | $10^2$–$10^3$ | 0 | Clostridia in 1 sample |
| Vegetable | $10^5$ | 0 | $10^3$ Clostridia in 1 sample |
| Laxative | | | |
| Vitamin Yeast | $10^1$–$10^4$ | 0–10 | Gram-negative rods in 3 samples |

cosmetic powders. Nowadays these powders are treated to reduce the microbial content so that the total number of viable bacteria and moulds is often less than 10/g and rarely exceeds a few hundred. There are exceptions, however, and examples of baby and wound dusting powders have been reported (Kallings *et al.*, 1966) to contain more than 1 000 bacteria/g, with coliforms and pseudomonads sometimes amongst those present. The fairly wide differences between the findings of various workers can only reflect the different qualities of the materials used and the method of treatment employed during manufacture, and they underline the importance of control of products of this type.

*Aqueous preparations.* One of the important findings in the Public Health Laboratory Service (PHLS) report (1971) was that heavy contamination (greater than 10 000 organisms/ml) occurs more frequently in aqueous than in oily or dry products and in general suspensions were more often contaminated than were solutions, ointments or creams. Of the solutions and suspensions peppermint water, magnesium trisilicate mixture, zinc sulphate lotion, kaolin mixture, expectorant mixture and ipecacuanhia and ammonia mixture were the worst offenders, the incidence of heavy contamination ranging from 70 per cent, in the first mentioned product to 49 per cent in the last mentioned. In general, products made in hospitals were more prone to contamination than were those obtained commercially.

The commonest contaminants found, especially in the alkaline or antacid mixtures, are the waterborne pseudomonads, sometimes *Ps.*

*aeruginosa*, and this indicates clearly that the primary source of contamination is the water supply. Growth of these bacteria can be quite rapid even though the pH value may be 9–10: in the absence of a preservative, counts of greater than $10^6$/ml have been recorded only a week or two after preparation.

Many aqueous solutions carry only a low level of contamination, depending largely on the solid ingredients used, and the examples quoted above refer to some of the worse offenders, but by no means exhausts the list. Syrups can also give rise to trouble, usually because of yeasts or moulds. The former are often of the osmophilic type and can result in gas production, with consequent explosion of their containers: the latter develop during storage and occur because of the evaporation of moisture from the syrup, the vapour condensing on the upper parts of the container and running back onto the surface of the syrup, thus providing a locally diluted sugar solution on which the mould grows with ease.

No problems arise with elixirs, linctuses and tinctures whose viable counts are nearly always below 100 organisms/ml.

*Creams and ointments.* Problems arise with these preparations because of their water content, and they arise from two aspects: (1) during manufacture and (2) during use. Water-in-oil preparations are less susceptible than are oil-in-water ones; anhydrous ointments do not permit microbial growth, but organisms can survive in them for long periods. It is not usual for the counts to rise as high as those in purely aqueous solutions, because of their entirely different rheological properties. For the same reason the contamination is very unevenly dispersed, growth occurring most readily at the surface where the oxygen tension is more suitable.

The types of organisms encountered are bacteria, moulds and yeasts, depending on the type of product and formulation. The bacteria include the waterborne *Pseudomonas–Achromobacter* types, the manufacturing contaminants, and the skinborne micrococci and staphylococci, the user contaminants. Moulds and yeasts of course, can be introduced at any stage. Several workers (*see* Sykes, 1969) have found creams containing corticosteroids to be particularly susceptible to contamination with *Ps. aeruginosa*. The present Oily Cream BP, which contains no preservative, and oily calamine cream are also susceptible to the same contaminants.

While hospital products are more frequently contaminated than those supplied by manufacturers, the major problem is from contamination which occurs during use in the ward or by the patient. The obvious way of overcoming this problem is by the use of suitable preservatives, but here again there are difficulties (see below).

*Eye, ear and nasal preparations.* Reference has been made already to the incidence of *Ps. aeruginosa* in eye drops with the consequent loss of sight in a number of cases, but this is not the only organism which has caused trouble. *Staph. aureus, B. subtilis* and *Proteus vulgaris* have also been incriminated, and not only in eye drops but also in lotions and irrigating solutions. All preparations used in, or coming into contact with the eye, including ointments, creams and contact lens solutions, are now classed as medicines and are required to be sterile; they should also be adequately preserved against subsequent contamination, unless dispensed in unit dose containers.

The same rigorous control is not required for ear and nasal drops and other preparations although they are also subject to similar contamination. Nevertheless, the Rosenheim Committee, in its final report (1973), recommended that 'the need for such a requirement (of sterility)' should be considered 'wherever appropriate'.

## Microbiological Standards

It is convenient to discuss the microbiology of medicinal products under two main headings: (1) non-sterile products and (2) sterile products. The standard in this context for finished sterile products is simply sterility, but the means for obtaining this are variable and complex: they are discussed later in this chapter. The standards for non-sterile products are much more a subject for discussion with particular reference to numbers and types of organisms permissible.

The most controversial question concerning non-sterile products is: should there be a numerical limit to the numbers of viable bacteria present? The Swedish authorities recommend that for all non-sterile medicinal products the viable bacteria count should not exceed 100/g or ml. If, however, the count is unavoidably above this level, as for example in preparations containing drugs of natural origin, they require checks to be made for the absence of certain specified bacteria. Other countries, notably Belgium and Denmark, follow the same line but do not specify such a high standard (or low count) for oral preparations. There is, in fact, a tendency in the Scandinavian countries towards requirements for sterility in preparations for application to delicate membranes and broken skin. In the United States Pharmacopoeia (USP); XIX (1975) maximum permitted viable counts are included in the monographs for gelatin and alumina and magnesia suspensions, the limits being 1 000 and 100/ml, respectively. In earlier editions of the USP there was a standard for Dried Yeast of not more than 7 500 bacteria and 50 moulds/g, but this has now been dropped. The Czechoslovakian Pharmacopoeia, second edition, also places a limit on tablets and other medicaments of 50 000 bacteria/g or ml, and the International Pharmacopoeia, second edition, (1967) states that 'microbiological contamination should be no greater than that permitted for foods under the law of the country concerned'; this is a broad specification in terms of total numbers of viable organisms but in some countries it could impose limits on the presence of certain types of organism.

In Britain the general consensus is against bacterial counts being included in official standards, on the grounds that they are notoriously variable and unreliable because: (1) they vary with temperature and other cultural conditions under which the tests are made; (2) in liquid preparations, the count is a varying factor depending on the age of the sample and its conditions of storage; (3) in solids, a uniformity of distribution of organisms is falsely and unjustifiably assumed and (4) unless a medicine is heavily contaminated the number of organisms consumed in a single dose would be unlikely to exceed the number consumed with, say, a glass of water or milk. Under (1) above, experience has shown that counts made on samples of liquids and solids in different laboratories under ostensibly the same conditions can differ by a factor of ten or more, and this alone militates against a numerical standard: the influence of temperature and medium on the growth of different types of bacteria does not need to be emphasised.

Because of these considerations the British authorities are much more favourably inclined towards control of specified pathogens or potential pathogens. At the same time they recognise the value of viable counts for 'in house' or process control by individual manufacturers, primarily as a means of assessing the maintenance of their own working standards and of adequate hygienic conditions during manufacture.

The pathogens of greatest significance are *Staph. aureus* (although this seems to occur only rarely in medicinal products), *E. coli*, the salmonellae and the pseudomonads, especially *Ps. aeruginosa*: consideration should also be given to the possible occurrence of clostridia in preparations which might be applied to small wounds and abraded skin. The British Pharmacopoeia 1973 now requires that *E. coli* and *Ps. aeruginosa* shall be absent from 1 g or ml of a few selected substances and formulations and that salmonellae are absent from 10 g or ml. Some workers consider, with some justification, that all pseudomonads should be controlled.

The materials at present thought to require control are those of natural origin plus a few liquid and solid preparations known to be susceptible to bacterial contamination and growth. The former include gelatin, tragacanth, thyroid powder, senna, digitalis and carmine; the latter comprise the alkaline and antacid mixtures with magnesium tri-silicate and kaolin/morphine mixtures. In addition the PHLS report (1971) and the Pharmaceutical Society's Working Party (*Report*, 1971) reveal high bacterial counts in peppermint and other waters, as well as in aqueous chlorhexidine and hexachlorophane solutions, syrups and mucilages, some dye solutions, for example, amaranth and tartrazine, and sulphadimidine and other mixtures. The BP standards for the absence of specific organisms are much in line with those of the USP which require salmonellae to be absent from acacia, charcoal, agar, digitalis powder and thyroid powder, and *E. coli* to be absent from gelatin and alumina and magnesia preparations; it is interesting that peppermint water has now been deleted from the USP.

In the hospital environment the primary and maybe exclusive concern is with pathogens. But what is a pathogen? A study of the literature shows that many organisms not normally classed as pathogens have been the cause of death of at least one subject, often an infant, which has little or no built-in or acquired immunity to infection of any sort, or a sick person whose natural resistances are at a low ebb. Examples of such organisms are some *Proteus* and *Flavobacterium* spp, *B. subtilis*, *Str. faecalis* and occasionally moulds, and one has to consider that in the hospital environment pathogens are likely to be much commoner than in the industrial or domestic environment.

Besides these pathogenic and opportunist organisms a further large group of bacteria, moulds and yeasts can cause sufficient spoilage of a wide range of products to render them unacceptable. They can cause changes in potency of an active ingredient, changes in odour, colour, taste and texture, breaking of emulsions, changes in viscosity, and gas production. Such changes are brought about by organisms using a specific drug as its source of carbon or metabolizing it in some way; by organisms producing destructive enzymes, e.g. penicillinases; by organisms metabolising ingredients in the product to give rise to 'fishy' taints, offensive fatty acids, hydrogen sulphide and other sulphur-containing compounds, as well as alcohols, hydrogen, carbon dioxide, esters, ketones, aldehydes and, of course, the development of a pigment or turbidity, with or without sediment, pellicle or scum. Some of the more exotic examples, as instanced by Smart and Spooner (1972) in their competent review of the subject, include the use of aspirin as sole carbon source by an *Acinetobacter* sp., the metabolizing of certain analgesics by a *Corynebacterium* sp., the metabolizing of atropine by *Coryne-bacterium* and *Pseudomonas* spp., the use of streptomycin as the sole carbon source for the growth of an *E. coli* subspecies and the use of phenols for the growth of *Pseudomonas* spp. Perhaps one of the most interesting examples is the growth of an anaerobic sulphate-reducing bacterium following a reduction in the redox potential of a suspension by aerobic bacterial growth, resulting in the production of hydrogen sulphide and the consequent blackening of the product due to traces of iron compounds being present. The ability of some actinomycetes to taint water with a musty or phenolic smell can also be significant in the hospital environment. This situation only arises when water is stored for long periods in tanks which are not cleaned out at sufficiently frequent intervals.

Adaptability to adverse conditions is another feature, some organisms and moulds have been observed growing on liquorice sticks, in Benedict's solution (which contains 1·6 per cent copper sulphate) and in normal sulphuric acid. But perhaps the most important, from the hospital standpoint, is the development of resistance in some Gram-negative bacteria, notably those of the waterborne types, to antimicrobial agents with selective actions such as cetrimide, chlorhexidine and the chloroxylenols. An interesting example occurred with one such organism which was 'trained' through several generations in progressively increasing concentrations of a chloroxylenol-containing product so that it could ultimately survive in a 20 per cent concentration of the chemical.


## Numerical Standards

The various compendia and other official publications specify, as already stated, a maximum permissible total viable count and/or the absence of named organisms from given amounts of the test material. But what do these mean in practice? In terms of total count, does it mean the highest number detectable under any test condition or the number detectable in one particular laboratory using its own current test conditions? The numbers could be quite different. For example, the USP XVIII says that the plate or multiple tube-MPN (most probable number) method can be used and that incubation should be for 48–72 h, but gives no temperature (except that for gelatin it specifies 30–35°C (86–95°F)). Sometimes even small differences in temperature, well within the range just quoted, can be significant in the recovery and growth of an organism, and, in terms of mixed populations such as are being considered, in the relative growths of the different organisms. Again, even if a specified culture medium is used, and this is often restrictive and inconvenient, batch variations occur in the medium, as well as slight differences in methods of handling in different laboratories, resulting in differing final counts. Finally, reference has been made already to the changes in viable population, both qualitative and quantitative, which occur, especially in liquids, during storage of a preparation.

All these considerations militate against establishing a numerical limit for the total viable microbial count in any pharmaceutical preparation and so support the general British attitude. They do not preclude its use, however, for comparison and in-process control in a single laboratory or manufacturing unit, for here the test conditions are likely to be subject to much less variation. It is in this field that the viable count has its place.

In terms of a specified or named organism, the statement that it must be absent from a given amount of the material under test is equivocal. For example, if *E. coli* is found in a single 1 g sample, when the requirement is that it be absent from that amount, the material is condemned, at least in theory. But is this correct? It is possible that the cell giving rise to the detection of the contaminant was the only one in the sample tested or even in the whole batch of material of which this was a sample. The requirement, therefore, is best considered as a statistical requirement, and only when looked at from this angle does it become reasonable. Accepting this, it seems that the 95 per cent probability level ($P = 95$), as is commonly used in other biological assays, gives a suitable standard.

## Specified Organisms and their Detection

In this direction, interest is wholly concentrated on contamination by pathogenic bacteria, the genera or species named being *E. coli*, the salmonellae, *Staph. aureus* and *Ps. aeruginosa*. As stated earlier, cogent reasons have been put forward to include all of the pseudomonads, but these have not yet been accepted. The clostridia merit consideration in some preparations, such as baby and wound powders, and mention has been made of *Candida albicans* and *Proteus* spp. No mention is made of viruses, on the basis that they are not likely to be present in isolation but always in association with bacteria.

*Methods of detection.* Details of the media or methods used for detecting the various organisms are given in compendia such as the BP 1973 and USP XIX (1975) as well as journals and textbooks on bacteriology. It will suffice here, therefore, to say that the first stage should always be an enrichment to increase the total number of bacteria present so that better growth is obtained in the subsequent selective medium. The recommended method of enrichment is by incubating the test sample for a short period, 4–18 h, in a liquid nutrient medium such as tryptone-soya broth: because only pathogens are under consideration incubation is usually at 37°C (98·6°F). (*See* Appendix A).

*Ps. aeruginosa* and *Staph. aureus* are detected by plating the enriched culture onto cetrimide agar and Vogel-Johnson agar, respectively, picking off appropriate colonies and checking them first morphologically and then by appropriate confirmatory tests. *E. coli* and the salmonellae are detected by culturing the enriched culture into a liquid lactose medium and, after 24 h, subculturing to: (a) selenite and tetra-thionate broths for the salmonellae, with subsequent plating on brilliant green and bismuth sulphite agars; and (b) McConkey agar for *E. coli*. In both cases, confirmatory tests are made on suspect colonies.

## Sterile Products

The manufacture and control of sterile products require special consider-ations because not only must the formulation be correct chemically but also the finished product must be sterile, and for this there are no tolerances; sterility is an absolute state. Moreover, just as chemical quality cannot be tested into any type of product, so sterility cannot be tested into a sterile product. Both have to be built in during manufacture to give the necessary quality assurance in the finished product, and testing is only the final check that the specified standards have been reached. These points are underlined in the revised Appendix (1974) to the GMP Guide (1971) on the manufacture of sterile medicinal pro-ducts. Some of the conditions specified in the Appendix to the Guide may seem to be over-stringent, but here one is dealing with products most of which are injected into the bloodstream and therefore by-pass the body's first line of defence: one must also remember that they are administered often when the normal body resistance is lowered.

Experience has shown that with such products even a small lapse in standards can have disastrous results, hence the need for extreme pre-cautions at all stages. Moreover the provision of correct premises, environment and other facilities impresses on the workers concerned the high standards expected and so gives the right setting for the quality of operations expected.

Quality in a finished sterile product can be considered in terms of: (1) premises and environment in which the work is carried out; (2) the equipment used, with emphasis on the ease with which it can be cleaned and sterilized, (3) the processes used, with particular reference to the methods of sterilization; (4) the personnel and (5) the controls employed during manufacture and in final testing. For products to be terminally sterilized, that is, processed cleanly and sterilized in their final containers, 'qualitatively similar but quantitatively lower standards' (GMP Guide, 1971) are permitted compared with those which are aseptically handled, that is, sterilized at some early stage and thereafter processed and dispensed aseptically.

## Premises and Environment

Essentially, a sterile product manufacturing suite consists of four areas or units:

(1). The preparation area for materials and containers.
(2). The clean or aseptic handling and filling areas.
(3). The changing rooms.
(4). The finishing area, which includes (for terminally sterilized products) the sterilizing area.

They should be arranged and designed to give a good workflow with no overlapping of activities. The finish of the floors, walls and ceiling should be of an impermeable material able to withstand regular washing and disinfection. There should be no cupboards, drawers or lockers (but the essential minimum of shelves may be permitted) and all windows should be sealed. Corners should be coved, there should be no bare wood or other materials which could give rise to dust and microbial contamination, and the equipment should be kept to the minimum required for the activities to be undertaken. Drains, sinks, ledges, pipes and similar structures are unacceptable in aseptic areas and undesirable in clean areas because they are notoriously prolific sources of bacterial contamination. Where sinks and drains must be installed in processing and changing areas effective traps which can be easily cleaned and disinfected should be fitted.

Protection against microbial contamination is especially important when carrying out aseptic operations, and protection against particulate contamination is essential. In the preparation of large volume intravenous and similar fluids special precautions are needed to reduce such risks to a minimum.

It is now the accepted principle that all clean and aseptic operations should be carried out in rooms or cubicles with controlled environmental conditions, i.e. flushed with filtered air, and that the immediate working area, the zone of greatest risk to the product, should be protected against airborne contamination by means of appropriately designed cabinets or hoods flushed with sterile air in a unidirectional, or laminar, flow style. There are some who consider that a laminar flow or similar cabinet in a normal laboratory environment meets the need, and even a few who think that a simple 'glove box' is adequate, but experience shows that a greater assurance and safety is obtained with good back-up conditions, that is, when the cabinet is in clean and controlled surroundings as just described.

The cabinet itself can be of the horizontal or vertical flow type (the former is preferable) and it should be so designed that the air flow at the work point is as near unidirectional as possible, without eddies and ingress of external air. The activities within the cabinet should be so arranged that the air first passes over the work area then over the hands of the operator and finally out and past the operator—this is a cardinal rule.

In the electronics industries much use is made of laminar flow rooms, but they have their disadvantages and there is no real place for them in hospitals or other small scale manufacturing units. Much more suitable, and less expensive, is the traditional aseptic room with its air supply brought in at high level and extracted or allowed to escape at low level. The points of entry and extract should be such that the room is flushed as uniformly as possible. This is difficult to achieve: a surprising number of eddies and turbulences may develop in an apparently simple system. In this respect each room is an individual entity.

The air flow rate is clearly an important factor and for clean operations a change rate of at least 8 changes per hour with air filtered to remove particles down to 5 μm is recommended; for aseptic operations the rate should be at least 15 changes per hour with air filtered to remove particles down to 0·5 μm. These rates are calculated to maintain the integrity of the rooms concerned in terms of airborne contamination during use. It goes without saying that small positive pressures should be maintained in the rooms in relation to the immediate surroundings in order to prevent vacuum effects.

Access for personnel should be only via changing rooms. These rooms should be designed on the 'black-grey-white' principle according to which operators enter first the black area where outer clothing and shoes are removed. They then sit on a barrier or sill built across the room and swing the legs over into the grey area where the hands are washed and the hood and mask are put on. In the white area the sterile suit is put on, hands are finally washed and sterile gloves put on. Where only terminally sterilized products are made a simpler 'black-white' system can be used, and clean instead of sterile dress is acceptable.

The whole changing area must obviously be flushed with a controlled air supply to prevent the build-up of airborne contamination which in such an area will be high. The direction of the air flow should always be from the white to the black zone, thus ensuring that any contamination generated in the white area is quickly swept away and that the heavier contamination expected in the black area is not allowed to diffuse back into the white area. A case could be made here for the use of a laminar flow type room.

## Personnel

No matter how carefully the premises are designed, the operator still remains the constant and major source of particulate and microbial contamination arising from both skin and clothing. The safest procedure is to isolate the operator completely from the surroundings by the use of a 'space suit' similar to that developed by Charnley & Eftekhar (1969) for hip-replacement operations, but this has its disadvantages.

The first requirement for all operators in a sterile product manufacturing area is that they shall have a high standard of personal hygiene. They must not be carriers of enteric pathogens and must not disseminate unduly large numbers of bacteria, especially *Staph. aureus*. Training is another important feature, and it should always be directed towards giving an aseptic sense so that the operator does not do anything which hazards the safety of the product.

It is well known that more organisms are shed from below than from above the waist and that skirts allow more organisms to be released than do trousers. It is recommended, therefore, that all operators should wear either one or two piece trouser suits with the neck, wrists and ankles overlapping the headgear, gloves and footwear, respectively, to minimize the number of skin particles disseminated into the atmosphere whilst the operator is moving around. The hair is a prolific source of bacteria and particles, hence a good head-cover is essential, preferably a full hood finishing over the shoulders. Similarly, footwear is an obvious source of micro-organisms, particularly of soil-borne sporing types, so that a change of footwear or at least a protective overshoe is imperative. Finally no comment need be made on the necessity of wearing gloves, especially when handling sterile materials.

## Sterilization

Sterilization can be effective with moist heat, dry heat, ethylene oxide, irradiation with high energy ionizing beams (X-rays or gamma rays), (but not with low energy ultraviolet (u.v.) radiation) and by filtration. The basic philosophy is that all of the material to be sterilized shall be subjected to the required intensity of treatment, whether it be temperature, gas concentration or radiation dose, for the prescribed period of time under the prescribed conditions. To do this the equipment must be correctly designed and the procedures carefully drawn up, and there must be adequate proofs and records of the efficacy of the treatment under practical working conditions.

Moist heat sterilization with steam at temperatures greater than 100°C (212°F) is the commonest and most reliable method and where appropriate should always be the method of choice. In hospitals the heat-plus-bactericide process is often used, especially for eye drops, but it is far less reliable than the normal steam pressure method.

Ultraviolet irradiation, though having bactericidal properties, is not a sterilizing agent: it has a limited range of action, its efficacy being related inversely to more than the square of the distance from the source and, being a low energy radiation, it has very low powers of penetration, even into liquids, including water. It is only effective against organisms exposed directly to the radiating beam. In spite of these limitations it can control within limits the numbers of viable airborne organisms: it is also useful in water treatment provided the right type of equipment is used.

*Steam.* It used to be a popular belief that no special skill was needed to operate a steam sterilizer and that almost anyone could handle the process satisfactorily. But experience during the last few years, including the unfortunate Devonport Hospital event (*Command*, 1972), has given the lie to this belief and shown that sterilization is a scientific process needing control and safeguards at every stage. Nowadays, the sterilizing cycle can be fully automated, but this does not mean that any of the controls can be relaxed, in fact, they may need to be watched more carefully. Permanent time-temperature record charts are now accepted as part of the control system of each manufactured batch.

One factor which often seems to be overlooked is the warm-up time, that is the time taken for individual items of a load to reach the required sterilizing temperature. The delay depends on the type and size of the individual items and of the total load and lags of up to 30 min can be expected between the time taken for the chamber to reach the sterilizing temperature and that taken for the contents of a 500 ml bottle to reach the same temperature (for more details, *see* Sykes, 1965). These delays must be taken into account in determining the duration of each stage of the sterilizing cycle and this can only be done by continuously comparing the temperature inside a typical container of a typical load against that in the chamber. Each type of container and each size of load requires individual consideration.

The chamber temperature is usually recorded at the coolest part of the chamber, namely, the drain, but in modern sterilizers the more normal procedure is to record the heating and cooling parts of the cycle from a separate thermocouple within the load or from a suitable simulator, and to control the actual sterilizing time from the chamber drain.

The treatments recommended in the compendia are heating for 30 min at 115°C (239°F) (10 lb steam pressure/in$^2$) or 15 min at 121°C (249·8°F) (15 lb steam pressure/in$^2$). For dressings and instruments a much shorter heating period at a higher temperature is used, namely, 3 min at 134°C (273°F) (30 lb steam pressure/in$^2$): rubber gloves can also be treated in a similar way, the important factor in their resistance to heat being the removal of oxygen. It has been suggested that the treatment at 115°C is not as effective as that at 121°C. This may be true under certain circumstances when large numbers of highly resistant spores are present but under normal conditions the treatment is perfectly satisfactory.

Much could be written on the types of containers and closures used for large volume intravenous and other fluids and the integrity of their seals. Suffice it to say here that only those sealed by fusion can be guaranteed, all others, involving separate closures, being fallible to a greater or less extent in that they are subject to recontamination from external sources after sterilization. The period of greatest risk is during cooling and often it is exacerbated by the use of rapid spraycooling, the cooling water itself being the source of the contaminants unless it is suitably pre-treated (Phillips et al., 1972). The commonest treatment is to keep the water at or above 70°F), thus maintaining it free from water-borne and other vegetative organisms.

*The heat-plus-bactericide process.* This process was introduced into the BP some years ago to meet the need for a simple sterilization process for substances in aqueous solution which cannot stand the normal autoclave treatment. It depends on the increased activity of a bactericide at elevated temperature but the bactericides permitted in the BP 1973 are limited to chlorocresol and phenylmercuric nitrate, both at twice the normal preservative concentration, namely, 0·2 per cent and 0·002 per cent, respectively. The process, which consists of heating the solution in its final container at 98–100°C (208·4–212°F) for 30 min, has severe limitations because it can only deal with small numbers of resistant cells. Its use in industry is small and in hospitals it is largely confined to the preparation of eye drops.

*Dry heat.* The mechanism of sterilization by dry heat is different from that of moist heat and consequently the conditions are different. A higher temperature for a longer period is needed and a minimum of 150°C for at least one hour is recommended. As with steam sterilization it is essential that the whole of the material to be sterilized receives this treatment. In other countries 160°C (320°F), or even 180°C (356°F), is specified but long experience in pharmaceutical manufacture has proved the efficacy of 150°C, mainly on the basis that all pharmaceutical materials are relatively clean microbiologically. Temperatures below 150°C for longer periods, e.g. 140°C for 4 h, have been used but they would only be acceptable for materials which are lightly contaminated.

To ensure uniformity of treatment an oven conforming to British Standard 3421 : 1961 should be used. This gives guidance on the size of oven, the insulation, the positioning of the heating elements and the use of fans for distributing the heat. Heating takes place mainly by radiation and secondarily by conduction, and this means that the oven must be carefully loaded and not overloaded. Consideration also needs to be given to the ability of the individual items or packages to allow heat penetration: in this respect powders present the greatest problems and should only be treated in thin layers. Again, as with steam sterilization, due attention must be given to the time taken to heat the load.

Dry heat is used for sterilizing glassware, instruments, non-volatile oils and waxes, soft or hard paraffins, glycerol and sometimes impregnated dressings. Syringes and other glassware can also be sterilized in an infrared heating tunnel. In this the items are passed individually on a continuous belt through the tunnel, part of which has heating elements so that the temperature of the items, which must be small, reaches 180°C and is held at this temperature for at least 20 min during its passage through the tunnel. Clearly the post-heating portion of the tunnel and the environment in which it is installed must be suitable for receiving and handling the sterilized items.

*Ethylene oxide.* Ethylene oxide is a highly reactive, toxic and explosive compound which boils at about 11°C (51·8°F). Sterilization with ethylene oxide is difficult to control and should only be undertaken by staff who are aware of the problems and who have full bacteriological testing facilities available. The efficacy of the process depends on the concentration of gaseous ethylene oxide present and on the temperature and humidity of the system. Treatments with 10 per cent gaseous ethylene oxide (200 mg/l) at not less than 20°C (68°F) for 18 h to 50 per cent gaseous ethylene oxide (1000 mg/l) at 60°C (140°F) for 3 h are used. In considering humidity conditions it is the state of hydration of the individual cells which is important and it may take some time before dried cells reach a state of equilibrium with the immediate micro-environment; the same difficulty is not experienced with partly hydrated cells. The optimum relative humidity is about 50 per cent; below 30 per cent the activity falls away rapidly.

Several factors besides those already mentioned need careful attention. First, to be effective the vapour must have direct access to the individual cells. Those occluded in crystals or protected by thin coatings of organic material such as mucus or even nutrient broth are immune from attack. Secondly, the vapour must be uniformly distributed within the packages of material to be sterilized, and fortunately in this respect ethylene oxide is readily diffusible. Thirdly, ethylene oxide is soluble to varying degrees in many types of material including rubber, most plastics, paper and board, as well as water with which it is miscible. It also hydrolyses in water to form ethylene glycol but if chlorides are present the products are the highly toxic chlorhydrins. Fourthly, because of its solubility in rubber and plastics, due time, running to several days, should be allowed after sterilization for de-gassing, so that the dissolved ethylene oxide is reduced to a safe level.

In spite of these difficulties ethylene oxide has a place in the sterilization armoury, its main value being for treating materials which cannot be sterilized by heat. It is used for sterilizing glass and plastic containers, rubber closures, plastic tubing, including giving sets, syringes and other equipment. It has a limited usefulness for sterilizing powders, mainly because of its high chemical reactivity.

*Irradiation.* Ionizing radiations generated from X-rays, cathode rays in an electrode accelerator or gamma rays from Cobalt-60 are effective sterilizing agents because of their high electron energy and powers of penetration. The capital cost for a radiation unit is high, largely because of the screening necessary to contain the radiation and protect those working near the unit. The accepted sterilizing dose is 2·5 Mrad although in Denmark 4·5 Mrad are required. This is because two resistant types of bacteria have been reported—*Str. faecium* and *Micrococcus radiodurans* —which, if present in sufficient numbers, can withstand a lower dose. These are exceptions; all other organisms are susceptible so that the lower 2·5 Mrad dose will deal with them satisfactorily provided the contamination level is not high.

Special equipment and premises are required to ensure that each item to be sterilized receives a uniform radiation dose. Consideration must also be given to the transparency of the materials to radiation, heavy metals being the most opaque.

The lethal action of radiations arises from their exciting effect on molecular structure and its consequent disruption, but this action is not confined to micro-organisms; there is no selective action, therefore the usefulness of the irradiation in pharmaceutical practice is severely limited. Its main value is with plastic syringes and other equipment; only a few medicinal substances can withstand the treatment without some degradation.

*Filtration.* This is probably the most exacting of the sterilization procedures because it first requires care in carrying out the actual filtration of the solution in bulk and then skill in the subsequent transfer to suitable containers, all of which must be done under strictly aseptic conditions. The mechanism of filtration is primarily by physical retention but adsorption can play a significant role, depending on the nature of the filter system.

The most widely used filters are membranes of cellulose acetate and other similar plastics. They act entirely as physical barriers and so the pores must be capable of preventing the passage of the smallest microbial cell—viruses present different problems and are excluded from filtration specifications. Other filters, such as those made from sintered glass or unglazed porcelain, depend partly on physical retention, partly on adsorption and partly on entrapment in the tortuous pathways through the depth of the filter. The asbestos pad filters, having no porosity as such, depend entirely on adsorption and entrapment. Although efficient when the right grade is used the asbestos filters have two main disadvantages: (1) they shed fibres which must be removed by means of a post-clarifying filter because of their danger when injected and (2) some drugs are absorbed by them.

In specifying porosities one must always consider the maximum pore size, not the average. For membrane filters $0.2$ µm pores are required; some workers use only $0.45$ µm but certain microbial forms can pass this porosity. For the depth filters, the maximum permitted pore size depends on the nature of the filter material and can only be determined experimentally. The reliability of asbestos pad filters must also be determined experimentally. These determinations are made by filtering a suitable volume of a suspension of test organisms ($10^7$–$10^8$ cells/ml) in a nutrient medium and incubating the filtrate. The recommended organisms are *Chromobacterium prodigiosum* or a selected small-celled *Pseudomonas* spp.

Filtration is used mainly for aqueous solutions, although it can also be applied to oily solutions. Its use should be restricted to those solutions which cannot withstand a heat-sterilization process.

## Process Monitoring and Control

Although one of the standard accepted methods of sterilization may be chosen, its efficacy must be proved in the equipment to be used and with typical loads of the material to be sterilized. Each piece of equipment and each sterilizer, even of apparently standard design, has its own performance characteristics. Similarly each type of material and each load, in terms of its size and packing, has its own responses to the treatment applied whether it be heat, gas or radiation. This monitoring is essential before any production sterilization is undertaken. It involves determining the physical parameters for each type of treatment and for each type of load, and translating these into standard procedures. These should be written as permanent references and no departure should be permitted unless it is specifically authorized. The complete cycle should be checked at regular intervals, including the recording instruments, because of their tendency to drift. The monitoring requirements for sterilization by steam are set out in more detail on page 221.

## Chemical and Biological Indicators

Chemical and biological indicators are frequently used to control a heat sterilization treatment, and sometimes they are the only controls used. The Rosenheim Committee (1973) concluded, however,

> that chemical and biological indicators are of doubtful reliability and hence of secondary importance,

and continued

> the correct way to control a sterilization process . . . . is to ensure that these (known physical) requirements are met by the use of a sound method of physical measurement

The same argument can be applied to radiation sterilization.

The reason for these unequivocal statements is that biological indicators contain bacterial spores—*B. stearothermophilus* for sterilizations by steam or dry heat—dried onto strips of filter paper or similar material, and by their nature cannot be prepared with a controlled heat resistance. To be effective as a control the spores should succumb in sufficient numbers precisely at the end of the prescribed sterilization treatment, or very near to it, but insufficient is known on the conditions required to produce cultures of adequately reproducible resistance, let alone resistance at a predetermined level. It is useless to have a test piece with an endpoint response well away from the sterilization endpoint. In spite of this the American authorities favour biological indicators and for terminally sterilized products accept for sterility testing a smaller number of containers from each batch if such indicators are used.

Chemical indicators for heat sterilization depend on a temperature-time reaction, the endpoint of which is indicated by a colour change designed to coincide with the sterilization endpoint. Of the indicators at present available, some e.g., Browne's tubes and certain of the colour spot cards, approach this criterion but others, notably the autoclave tapes, can only indicate that the material has been subjected to a heat treatment. Ageing and conditions of storage have some bearing on the endpoint, and storage in the cold is recommended. The main value of chemical indicators is in revealing a failure to sterilise: a failure with the indicator is certain evidence of a failure of the treatment, but the converse is not necessarily correct.

The 'red perspex' or Perspex HX indicator used in controlling radiation sterilization is in a different category. It is an accurately devised test piece based quantitatively on a radiation sensitive reaction and so is not subject to external influences in the same way as are the other chemical indicators. It is therefore a far more reliable test piece and is far superior to any biological indicator.

The one method of sterilization for which there is no chemical substitute for the biological indicator is by ethylene oxide. Here the combination of gas concentration, time, temperature and humidity is not adaptable to physical or chemical measurements and so recourse has to be made to a biological agent. The indicator normally consists of the spores of a culture of *B. subtilis* var. *niger* (*globigii*) dried from an alcoholic suspension on to small pieces of aluminium foil (Royce's sachets). Various chemical test pieces have been tried but with very limited success.

## Tests on Finished Products

Besides the chemical tests normally applied to a manufactured medicinal product the pharmacopoeias and other guides specify tests for sterility and in some instances test for the absence of pyrogens as well as tests for the absence of particulate matter. Whilst most, if not all, agree that a chemical analysis on a finished product is essential, even though the quality and quantity of the ingredients may have been checked at each stage, there are those who do not accept the need for a check for sterility on products which have been sterilized by heat or radiation in their final containers; they prefer to rely wholly on the monitoring of the cycle and on the in-process controls. The Rosenheim Committee (1973) considered carefully the values of sterility testing and wrote

> sterility testing of a batch is a valuable safeguard capable of detecting contaminated samples . . . . for this reason the Committee support the reasoned application of the test as part of the control procedure.

The Appendix to the GMP Guide (1974) explains further that

> to omit the sterility test and so rely solely on in-process tests and other controls may lead to serious risks. Continued compliance with the sterility test gives increasing confidence in the efficacy of a sterilization or aseptic process.

*The test for sterility.* Testing for sterility, or checking for the absence of organisms, is a different process from that of looking for the presence of certain bacteria, and requires a different approach.

The test consists of examining samples from each batch of product for the presence of aerobic and anaerobic bacteria and fungi. A batch for this purpose consists of that number of containers prepared in such a manner that the risk of contamination is the same. For terminally sterilized products it is the number contained in a single sterilizer load; for aseptically filled products the number filled from the same bulk.

The numbers of containers to be sampled and the amounts to be tested from each container are specified in the European Pharmacopoeia Vol. II (1971). The numbers and amounts to be tested differ somewhat in other pharmacopoeias but it should be borne in mind that any quantity stated is a minimum. Larger volumes can be tested and, in fact, the Appendix to the GMP Guide (1974) indicates that for container volumes greater than 50 ml the whole content should be tested, this being in contrast to the European Pharmacopeia's recommendation of only 10 per cent of the volume.

Wherever possible the membrane filtration method should be used, thus minimizing the volumes of culture media and eliminating the carry-over into the test of antimicrobial preservatives or other agents. If membrane filtration is not possible, the required volume of sample is added to sufficient liquid nutrient medium containing an appropriate agent to neutralize the antimicrobial effect (See Table 7.2).

The media recommended are either a dual purpose aerobic-anaerobic one, e.g., liquid thioglycollate medium, or separate aerobic and anaerobic media, e.g., tryptone-soya broth and Robertson's meat medium, respectively. The medium for fungi is a Sabouraud-type liquid medium or a nutrient broth containing 1 per cent dextrose. The tests are incubated for seven days at 30–32°C (86–89·6°F) for bacteria (the Euro-

**Table 7.2.** *Inactivating agents for antibacterial substances.*

| Antibacterial substance | Inactivating agent |
|---|---|
| Quaternary ammonium compounds; certain guanadine derivatives and other compounds | Lecithin-Tween 80* |
| Quaternary ammonium compounds chlorohexidine | Lubrol W† |
| Mercury compounds | Cystein thioglycollate |
| Chlorine-releasing compounds | Thiosulphate |
| Formaldehyde | Sodium metabisulphite (dilution better?) |
| Chloramphenicol | Acetylating enzyme |
| Polymyxin, erythromycin | pH 6 |
| Streptomycin | semi-carbazide |
| Sulphonamides | *p*-aminobenzoic acid‡ |
| Trimethoprim | Thymidine |
| Penicillins | β-Lactamases |

\* 0·07% lecithin + 0·5% Tween 80
† 3·0% is recommended; not fully effective
‡ minced meat as in Robertson's medium is also effective
§ For substances with no known inactivator, reliance has to be placed on adequate dilution.

pean Pharmacopoeia states 30–35°C (86–95°F)) and at 22–25°C (71·6–77°F) for fungi. The normal temperature of 37°C (98·6F) is not used because damaged organisms generally recover better at the lower temperatures and the test is intended to detect not only potential pathogens but also as many other types of micro-organism as possible.

When there is growth in the first test, repeat tests are permitted under some circumstances but they should not be undertaken routinely; each should be considered on its merits and in the light of the type of organism found in the first test. It is not permitted to release the material for use until at least one test showing no growth has been recorded from amongst the first three tests made; it is not permitted to make more than two repeat tests. Because of the significance of growth in any test great care is needed to eliminate the hazards of accidental contamination from sources other than the sample itself. This means that all manipulation should be undertaken only by skilled operators working under strictly aseptic conditions. Failure to do this may lead to failure of otherwise satisfactory material; the cause of the failure should always be investigated to avoid, if possible, its recurrence and the findings recorded.

*The test for pyrogens.* All large volume sterile fluids for administration to patients are required to be pyrogen free. The only test at present approved for this purpose is based on the in vivo response of rabbits but the in vitro Limulus lysate test recently reported (Cooper et al., 1971; Eibert, 1972), is being actively investigated in a number of laboratories, particularly in the radiopharmacy field. Its present status is that both false positives and false negatives have been reported and some preparations have failed to show any response.

For the rabbit test details are laid down in the official compendia and the reader should consult one of these for further information. Each animal naturally shows a small positive temperature response according to the way it is handled and the official tests make allowance for this. Any temperature above this normal increase, but within the limit set in the official test, should be investigated because it could indicate the beginnings of a breakdown at some stage of the manufacturing procedure.

*Tests for particulate matter.* For obvious reasons all injections should be as free as possible from particulate matter. The BP 1968 says that they should not contain particles that 'can readily be observed on visual inspection'. In addition, the BP 1973, following the Australian precedent, places limits on the numbers of particles down to 5 µm and 2 µm in large volume injections.

Visible particles, most commonly fibrous, originate mainly from the containers and closures. Small particles are more elusive, and it is difficult to imagine how so many such particles can appear in a solution which has been filtered through a membrane of 0·22 or 0·45 µm porosity. Even worse, these particles are known to increase in number during transit and storage, the extent of the increase depending on the amount of agitation during transit and on the conditions during storage.

Examination for small particles is a specialized technique involving electronic counting equipment, of which the Coulter counter is an example, and can only be carried out by those skilled in handling the equipment. The Australians originally recommended a limit of not more than 500 particles per ml measured at 3·5 µm and this has been translated in the BP 1973 to not more than 100 particles at 5 µm and 1000 particles at 2 µm. These are averages for a number of containers from a batch; double these numbers are allowed for a single container.

Examination for larger, visible particles is a highly subjective operation particularly when one attempts to interpret in practice the term 'readily' as used in the BP 1968. Their detection depends to a great extent on a number of personal and physical characteristics, amongst which are the age and general well-being of the examiner and on the actual test conditions, especially the intensity and type of illumination used.

Because of the random nature of the distribution of visible particles every container in a batch or delivery should be examined before it is used.

A general concensus appears to be emerging that a large volume container should not be examined for too long a period, and 5–10 s is suggested as adequate. The best illumination is thought to be a 60 W tungsten lamp sited so that the particles are seen by reflected light against a dark background, the direct rays of the lamp being shaded from the eyes of the observer. Opinions about the relative value of polarised light differ widely. Its main advantage is that fibrous particles are revealed more clearly; on the other hand, white amorphous particles can become invisible in this light.

## Use of Antimicrobial Preservatives

Because of the susceptibilities of some medicinal products to microbial contamination and spoilage, appropriate steps must be taken to prevent such occurrences. Selection of microbiologically clean starting materials, if such a choice is possible, and strict attention to hygiene during manufacture will do much in this direction and may even eliminate it, therefore good housekeeping and good manufacturing practice are basic essentials. It is not unknown for contamination to be so light that the organisms are unable to develop, and can even die, whereas with heavier contamination the same organisms can grow rapidly. There is still the problem, however, of maintaining the integrity of the product during storage and use, and it is here that a preservative serves its main function. No argument can be sustained for the use of preservatives to mask bad manufacturing practice.

The antimicrobial agent used should be not only effective in preventing microbial growth but it should also not interfere with the therapeutic action of the drug used or produce turbidity; it should be free from toxicity and irritancy at the concentration used; it should be stable in the preparation and it should preferably be non-selective in its action. The list of compounds meeting these requirements is somewhat restricted and those most commonly used are chlorocresol, chlorbutol, cresol (phenol is no longer recommended), phenylmercuric nitrate (PMN) or acetate, the parabens (various esters of *p*-hydroxybenzoic acid, or Nipa products), cetrimide, chlorhexidine, benzalkonium chloride, benzyl alcohol and chloroform: acid pH values also discourage the growth of bacteria, but not yeasts or moulds, and the alcohol present in some formulations is useful, particularly against moulds. These agents are not interchangeable; each has its own characteristics, with its advantages and disadvantages, therefore each has its own range of application. It is not the intention here to discuss the properties of the preservatives named (those interested should consult *Disinfection and Sterilization* (Sykes, 1965)), except to note that in any preservative lethal rather than simply inhibitory properties are always looked for. In fact, the BP 1973 describes a test, intended primarily to assess bactericides used in parenteral preparations, in which selected non-sporing organisms are expected to be killed within three hours. This may be an unnecessarily short period for oral preparations, but the period should not be extended beyond about 24 h, otherwise adaptation and consequent growth of some of the surviving cells may occur. Somewhat in contrast, the USP XVIII describes a test in which chosen vegetative bacteria and fungi are inoculated into the preparation under examination and the mixture kept at 30–32°C, survivor counts being made at intervals during a 28 days test period. The preparation is considered to be satisfactorily preserved if 'the viable vegetative bacteria are reduced to not more than 0·1 per cent of the initial number and remain below that level for a seven day period within the 28 day test period': the fungi must not show 'any significant increase in numbers'. This appears to be too lenient a standard and does not take into account the risk of bacterial adaptation.

The activities of all antimicrobial agents are affected, usually adversely, but occasionally advantageously, by the pH value and by other ingredients of a medicinal product, therefore it is not sufficient to assess their action in simple aqueous solution. Each agent must be tested in the preparation itself and against the contaminants likely to be found in the preparation as well as against the organisms recommended in the official compendia, namely, strains of *Ps. aeruginosa, Staph. aureus, E. coli* and *C. albicans*. These restrictions tend to reduce the possible choice of substances of an already limited list.

This situation applies particularly when oily preparations, emulsions, creams and ointments are involved, for here there is the additional problem of the partitioning of the preservative between the oil and the aqueous phases. Only that proportion of the substance which remains in the aqueous phase can exert any preservative action. These systems are complex and have been studied in detail by Bean and his colleagues, to whose publications further reference should be made (see Bean et al., 1965; Bean et al., 1969).

*Sterile products.* Preservatives in solutions for injection are confined to those dispensed in multi-dose containers, though it is wise to put them in as an additional safeguard against contamination during aseptic filling. They are prohibited in preparations administered in volumes of

I

15 ml or greater and in certain other injections, depending on the route of administration.

As stated above any preservative used should fulfill the BP requirement of effective lethal action within three hours.

Rubber closures present a particular problem in terms of the stability of a product during storage, owing to the solubility of preservatives, notably the phenolic substances, in rubber. The problem is further accentuated if the substance is volatile, thereby being dissolved at the inner surface of the closure, diffusing through the rubber and being lost at the outer surface, or, if it is used at low concentration, e.g. phenylmercuric nitrate at 0·001 per cent, when a small loss can have a significant effect on the final concentration. To meet this the BP specifies a special pre-treatment of the closures in water containing twice the strength of the bactericide used in the injection. This does not meet the situation fully but at least it delays the loss of bactericide from the injection.

## Conclusion

The preparation of medicinal products is a dynamic process and frequently suggestions and recommendations are made to improve equipment, processes and standards, chemical, physical and microbiological. Provided they are not changes for changes' sake or standards for standards' sake they are good and can only lead to better quality assurance in the finished products. Both hospitals and industry have parts to play in these developments and those concerned should keep abreast of them.

## Appendix. Preparation of Samples for Testing

No problem arises in the preparation of aqueous solutions or suspensions for estimating the microbial content or for sterility testing, except that care should be taken with suspensions to make as even a dispersion of the solution as possible before removing the test sample. With oily preparations, ointments and creams some preliminary preparation is necessary before the culture work is undertaken.

No organisms will grow in oily preparations in the absence of water, but they can survive for long periods. To detect such organisms they must be induced to move from the oily to the aqueous phase and this is done most easily by shaking the oily sample with a large volume (at least 10:1 v/v) of an aqueous medium to give a relatively fine dispersal of the oil. Water and saline do not give as good recoveries of organisms as do quarter-strength Ringer solution or 1 per cent peptone water, preferably containing a small amount (up to 0·2 per cent) of Tween 80; nutrient broth is also satisfactory. Having made the aqueous 'extract' the standard testing procedures can then be followed, remembering also that any antibacterial agent may also need to be neutralised or inactivated (Table 7.2).

The size of the sample can have a bearing on the numbers of organisms recovered, small amounts up to 1 g or ml giving better recoveries than larger amounts of, say, 10 g or ml, presumably because the smaller amounts can be more easily dispersed and, therefore, 'extracted'. Several publications have recommended membrane filtration techniques, including the use of diluent-dispersing agents, such as isopropylmyristate, when the original material is viscous. The membrane is then cultured direct, or washed with an inert organic solvent, and then cultured. There are difficulties, however: (1) membrane filtration of any oil is a slow process; (2) clogging of the membrane can easily occur, especially with isopropyl myristate; (3) no organic solvent can be guaranteed to be free from antimicrobial action, and (4) the whole procedure involves additional manipulative stages with the consequent hazard of accidental contamination.

Ointments must be diluted with a low viscosity oil to give them greater fluidity, and so render them more easily extractable, otherwise the sample will float in the culture medium. The pharmacopoeias recommend that this should be done at a temperature not exceeding 45°C (113°F) (to soften the ointment), the mixture being shaken or stirred for only a few minutes, but even this mild treatment could be lethal to some bacterial cells, so that care must be taken in following this method.

Oil-in-water creams are not difficult to handle because the continuous phase is water, rendering the organisms easy to extract into the larger volume of diluent. Water-in-oil creams, however, are different and should be treated in the same way as oily preparations, described above.

## References

Ayliffe, G. J. A., Barry, D. R., Lowbury, E. J. L., Roper-Hall, M. J. & Walker, M. W. (1966) Post-operative infection with *Pseudomonas aeruginosa* in an eye hospital. *Lancet*, **1**, 1113.

Bean, H. S., Heman-Akha, S. M. & Thomas, T. J. (1965) The activities of antibacterials in two-phase systems. *J. Soc. Cosmet. Chem.*, **16**, 15.

Bean, H. S., Korning, G. H. & Malcolm, S. A. (1969) A model for the influence of emulsion formation on the activity of phenolic preservatives. *J. Pharm. Pharmac.* **21**, 193S.

British Pharmacopoeia (1968) London: General Medical Council.

British Pharmacopoeia (1973) London: H.M.S.O.

British Pharmaceutical Codex (1973) London: Pharmaceutical Press.

Charnley, J. & Eftekhar, N. (1969). Post operative infection in total prosthetic replacement arthroplasty of the hip joint—with special reference to the bacterial content of the air of the operating room. *Br. J. Surg.*, **56**, 641.

*Command* (1972) Report of the Committee appointed to enquire into the circumstances, including the production, which led to the use of contaminated infusion fluids in the Devonport section of the Plymouth General Hospital. London: H.M.S.O.

Cooper, J. F., Levin, J. & Wagner, H. N. (1971) Quantitative comparison of *in vitro* and *in vivo* methods for the detection of endotoxins. *J. lab. clin. Med.*, **78**, 138.

Eibert, J. (1972) *Bull. Drug. Assn.*, **26**, 253.

European Pharmacopoeia (1971) II. France: Maisonneuve.

Guide to Good Pharmaceutical Manufacturing Practice (1971) with Appendix 1 (Revised) Sterile Medicinal Products (1974) London: H.M.S.O.

Kallings, L. O., Ernerfeldt, F. & Silverstolpe, L. (1965) The 1964 Inquiry by the Royal Swedish Medical Board into Microbial Contamination of Medical Preparations: Final Report.

Kallings, L. O., Ringertz, O., Silverstolpe, L. & Ernerfeldt, F. (1966) Microbiological contamination of medical preparations. *Acta Pharm. Suecica*, **3**, 249.

International Pharmacopoeia (1967) 2nd edition, W.H.O. Geneva.

Medicines Act (1968) London: H.M.S.O.

Pedersen, E. A. & Ulrich, K. (1968). Microbial content of non-sterile pharmaceuticals III. Raw materials *Dansk. Tidss. Farm.* **42**, 71.

Public Health Laboratory Service Working Party Report (1971) Microbial contamination of medicines administered to hospital patients, *Pharm. J.*, **207**, 91.

Phillips, I., Eykyn, S. & Lake, M. (1972) Outbreak of hospital infection caused by contaminated autoclaved fluids. *Lancet*, **1**, 1258–1260.

Pyrogen testing: horseshoe crab versus rabbits.

Report: Microbial Contamination in Pharmaceuticals for Oral and Topical Use (1971) *Pharm. J.* **207**, 400.

Rosenheim Committee (1973) *Report on the Prevention of Microbial Contamination in Medical Products.* London: H.M.S.O.

Smart, R. & Spooner, D. F. (1972) Microbiological spoilage in pharmaceuticals and cosmetics. *J. Soc. Cosmet. Chem.*, **23**, 721.

Sykes, G. (1965) *Disinfection and Sterilization*, 2nd edition. London: Chapman & Hall.

Sykes, G. (1969) Microbial contamination in pharmaceutical preparations for oral and topical use: a review. *Indian J. Pharm.*, **31**, 33.

Sykes, G. (1971) Control of microbial contamination in pharmaceutical products for oral and topical use—raw materials. *J. mond. Pharm.*, **14**, 8.

United States Pharmacopoeia (USP) XVIII (1970) United States Pharmacopoeial Convention Inc., Washington DC.

United States Pharmacopoeia (USP) XIX (1975) United States Pharmacopoeial Convention Inc., Washington DC.

# 8 | Skin as a Microbial Habitat

## W. C. NOBLE

The skin has long been known as a source of micro-organisms—often merely as a source of contaminant micro-organisms—but the study of skin as an individual habitat, as an area worthy of investigation for its own intrinsic interest, is comparatively recent. The upsurge of interest which followed the publication of Mary J. Marples' book *Ecology of Human Skin* in 1965 was a tribute to her efforts and a reflection of the interest in opportunistic pathogens in disease processes of, for example, immunologically compromised patients (Savin & Noble, 1975). Out of this pre-occupation with skin organisms has arisen the study of cutaneous micro-ecology.

## Physical Features

The skin is the largest organ of the body, it weighs about 5 kg and has an area of about 1·75 m$^2$ in an average adult. It is differentiated into various structures such as nails and hair and has three types of gland which open onto its surface or whose products appear at the skin surface. The skin's physical features in relation to microbiology were reviewed by Noble and Somerville (1974).

The visible part of the skin surface is composed of flat, pavement like cells known as epithelial cells which are derived from a reproductive layer deeper in the skin. Cells from this basal layer gradually move towards the exterior of the body; they lose their cuboidal shape and become flattened, whilst doing so they become filled with a protein known as keratin. When they reach the surface they are no longer viable and are gradually lost. Loss of epithelial cells as 'skin scales' or 'squames' is a natural process and some 10$^7$ fragments are lost each day as a result of friction and washing; skin scales may also be lost from the surface in the absence of friction as a result of drying, when the cells curl up and peel away. Each individual squame is about $30 \times 30 \times 3–5\ \mu m$ and many dispersed fragments are small enough to remain airborne for long periods, having a settling rate in still air of less than 5 mm/s. Small wonder that they should be of importance in the dispersal of micro-organisms from the body surface.

There are differences in dispersal between males and females which are not yet fully understood. Males disperse more organisms than do females, they disperse specifically far more particles bearing *Staphylococcus aureus* and they appear to disperse from different body areas (Noble, 1975b). Dispersal of total body flora is most strongly correlated with organisms of the thigh and abdomen in males but with the shin in females. Some of these differences may be attributable to the heavier colonization of males with aerobic bacteria and others to traditional differences in clothing but much remains to be learned. There are differences in the size of epithelial cells over different areas of the body and on average the cells of females are larger than those of males (Plewig, 1970). However it is not yet possible to decide whether these differences are reflected in the microbiology of the sites.

The skin temperature is about 33°C (91°F) though there are some marked regional differences. The feet are generally the coldest body area whilst the axillae have a temperature which approximates to that of the interior of the body. Except in old age skin is maintained at a surprisingly constant temperature even in adverse environmental conditions.

Skin pH is about 5·5 though the normal range is from 4·0–7·5; there are differences between body areas though these are less than differences between individuals. It has been suggested that the lactate–lactic acid content of eccrine sweat acts as a buffer though other substances may contribute, $CO_2$ diffusing through the skin may serve to neutralize surface alkali. A comprehensive study of skin pH was published by Behrendt and Green (1971).

There seems little doubt that the factor with the greatest influence on the density of microbial populations on the skin is humidity. When the skin becomes dry Gram-negative bacilli, 'coliforms', are unable to survive (Payne, 1949) but when the skin is occluded populations of all organisms reach high levels, though there may be an overgrowth of a pathogen, resulting in lesion formation (Marples & Kligman, 1969; Chujoh, 1971; Knight, 1972; Singh, 1973, 1974; Jones et al., 1974).

According to the detailed studies of Duncan et al., (1969) and McBride et al., (1975) an increase in both temperature and humidity is necessary before skin populations increase and this point has often been overlooked in other studies. It is supported by the studies of Horne (1952) and Taplin and his colleagues (1965, 1973, 1974) on the relation of climate to skin disease. Aly and his colleagues (1975) have now attempted to synthesize in vivo and in vitro observations into a coherent story.

There are three types of gland whose secretions appear at the skin surface and which have a profound effect on the ability of the skin to support microbial growth. These are the eccrine glands, apocrine glands and sebaceous glands.

## Eccrine Sweat Glands

These are the common sweat glands and are found over the whole body surface, though there are marked regional differences in density and activity. On average there are over 120 eccrine sweat glands per square centimetre of surface with a range from 240 on the dorsum of the hand to $82/cm^2$ on the leg. There is an inverse relation between the density and the output per gland. Total sweat output is greater at the temples and midline of the chest, abdomen and back and is greater in males than in females. Glands in the palms and soles respond to nervous stimuli whilst those in other areas are thermally controlled. The secretion of these glands is composed mainly of water (99 per cent) and inorganic ions including chloride, phosphate, sulphate, bromide, fluoride, iodide, potassium, sodium, calcium, magnesium, iron, copper, manganese and zinc. However, much organic matter also appears in eccrine secretion and urea, thiourea, creatinine, uric acid, urocanic acid, acetyl choline, mucoprotein, glucose and other sugars, pyruvic acid, lactic acid and lactates have been identified; Kuno (1956) recorded 10 vitamins in sweat and the amino-acid composition includes 23 compounds of this type. Immunoglobulins may also be found in sweat (Cabau et al., 1974; Förström et al., 1975). Details may be found in Noble and Somerville (1974). In studies on sweat composition, it has frequently been the case that water soluble material from the skin surface has been described rather than sweat per se. Although there are some striking differences between sweat from patients with cystic fibrosis and sweat from normals, no attempt seems to have been made to relate this to microbiology. It may be however that this would prove a difficult exercise since the composition of sweat in normals alone is influenced by sweat rate, diet and site of sampling.

Much is known of the chemical composition of surface scale in relation to some skin diseases. In psoriasis, for example, there is more zinc and iron but less cobalt in the scale than in normal skin (Molin & Wester, 1973a, b, 1974). There is a decrease in cystine but an increase in taurine in psoriatic scale (Sobhanadri et al., 1974); phosphorus compounds are also present in greater quantity in diseased skin (Mier & McCabe, 1963). Other studies have identified altered protein composition of psoriatic scale (Neufahrt et al., 1975) and enzyme studies are in in progress (Hammar, 1975). No attempt has been made to link these differences to the microbial flora, though Kloos and Noble (unpublished) believe that genetic differences may exist between the organisms on psoriatic plaque and those on adjacent normal skin.

Except for the sugar content there is little work directly relating microbiology to water soluble nutrient at the skin surface. Smith (1971) has shown that members of the *Micrococcaceae* can use lactic acid as an energy source even in the presence of oleic and palmitic acids. Coutinho and Nutini (1963) have described the amino-acid requirements of various phage types of *Staph. aureus*. *Micrococcus* species are more versatile than *Staphylococcus* species in their requirement for amino acid (Halvorsan, 1972; Farrior & Kloos, 1975). The presence of free glucose at the skin surface encourages the growth of pathogenic fungi (Zaun & El Mozayan, 1973) but attempts to relate the diabetic state to the incidence of infection have been confusing and indeed conflicting. Savin (1974) has presented a comprehensive review of this topic finding contradictory evidence for each well established 'fact'. In a series of well controlled diabetic patients studied by Somerville and Lancaster-Smith (1973) only fluorescent diphtheroids were more common than in the normal population studied.

In a study of the materials available for microbial growth at the skin surface Murphy (1975) found only a few organisms (*Micrococcus* spp. and some diphtheroids) able to grow on artificial sweat. The medium was shown not to be inhibitory, but to lack some factor, as yet unidentified, which was available in samples of human and guinea pig epidermis but not in hair. The substance is water soluble, stable to heat and passes through dialysis membranes. It can be substituted for by Coenzyme A though this does not correspond in chromatograms to the stimulatory substance. Omoregha and Noble (unpublished observations) have confirmed Murphy's finding in a study of 1400 cocci isolated from normal skin; all of the *Micrococcus* species described by Kloos, Tornabene and Schleifer (1974) but only a few *Staph. epidermidis* and *Staph. saprophyticus* strains can grow in the absence of the stimulatory factor.

There is some evidence that sweat, when freshly excreted, may be toxic to skin bacteria (Noble, 1975a), this may be attributable to the lactic acid content.

Research in veterinary medicine promises to tell us much of conditions at the skin surface of other species (Jenkinson et al., 1974; Mabon, 1974).

## Apocrine Gland Secretion

The apocrine gland secretion has been less fully investigated than that from the eccrine gland and little is known of its relation to microbiology. Apocrine sweat, in contrast to eccrine sweat, contains protein and dries very quickly when exposed to air. Apocrine glands are found principally in the axillae and anogenital region and around the nipples. This distribution has led to suggestions that the secretion may be involved in the formation of body odour but this is not yet fully investigated. Modified apocrine glands are found in the ear where there are between 1000 and 2000 ceruminous glands which secrete an antimicrobial wax.

## Sebaceous Gland Secretion

Sebaceous glands are primarily found in hair follicles; though on the face, and especially the nose, the hair may be so reduced as to appear absent, the predominant feature being the sebaceous gland and its secretion. There may be over 500 sebaceous glands/cm$^2$ on the forehead and about 100/cm$^2$ on the rest of the body. Sebum is a complex mixture and, as measured at the skin surface, contains diglycerides, triglycerides, wax esters, squalene, cholesterol and cholesterol esters, and free fatty acids. Much of the free fatty acid is probably derived from hydrolysis of triglycerides by the local bacterial flora. Fatty acid composition is of great complexity; although there is a predominance of $C_{14}$, $C_{16}$ and $C_{18}$ chains, fatty acids may be found from $C_8$ to $C_{36}$, both straight and branched chains being represented; Kellum and Strangfeld (1972) detected 59 fatty acids in the range $C_8$ to $C_{18}$ alone.

The composition of the sebum changes in response to age, sex, physiological state and environment and there is a circadian rhythm of secretion (Burton et al., 1973; Noble & Somerville, 1974; Weirich & Longauer, 1974; Eberhardt, 1974; Downing et al., 1975; Gloor, 1975; Pablo & Fulton, 1975).

Skin surface lipids may be responsible for the suppression of potential skin pathogens such as streptococci and *Staph. aureus*. Aly et al. (1972) have shown that removal of fatty acid by acetone enhances the survival of staphylococci and streptococci on the skin and that replacement of the acetone soluble material restores the inhibitory action. Lacey (1968a, b, 1969) demonstrated that fat solvents remove the inhibitory factor as does neomycin which binds fatty acid. Short chain fatty acids ($C_8$) are inhibitory to *E. coli* even for those strains able to metabolize the longer ($C_{14}$) chained compounds (Weeks et al., 1969). Diphtheroids are inhibited by some fatty acids and stimulated by others (Pillsbury & Rebell, 1952; Puhvel & Reisner, 1970; Somerville, 1973).

There is some disagreement on the *requirement* for fatty acid described by some authors (e.g., Pollock et al., 1949; Smith 1969); Somerville (1973) found only a few diphtheroids unable to grow in defatted media, though many grew much better in the presence of fatty substances. *Pityrosporum ovale* appears to need fatty acids for growth, usually in the form of myristate ($C_{14}$) or palmitate ($C_{16}$) (Shifrine & Marr, 1963). Earlier workers who reported that oleic acid was essential may have been misled by the high content of myristic, palmitic and other fatty acids in commercial oleic acid.

Stimulation of skin bacteria by lipids has been studied using in the main such compounds as the Tweens. Tween 80 (polyoxyethylene sorbitan mono-oleate) is used most frequently and has been found to stimulate growth; it is attacked by many diphtheroids which possess a specific esterase whereas other lipid compounds are not (Smith & Willet, 1968).

There is scope for more experimental work on microbial lipases in relation to the skin flora. *Staph. epidermidis* strains have a wider range of lipolytic activity than do either lipophilic diphtheroids or strains of *Corynebacterium acnes* (Smith & Willett, 1968). In a search for enzyme inhibitors which might permit control of lipolysis in acne, Pablo and his colleagues (1974) found that whilst enzyme preparations from *Staph. epidermidis* split the α position of triglycerides more rapidly than the β, giving rise to monoglyceride intermediates, enzymes from *C. acnes* split both α and β equally producing glycerol and free fatty acids. Further, the *Staph. epidermidis* enzyme was active in the range pH 7–10 whilst that of *C. acnes* was active from pH 5–8. Since skin pH is between 5 and 6 this may explain the greater role of *C. acnes* in lysis of skin surface lipid (Marples et al., 1974) and in the causation of acne.

Puhvel and her colleagues have made elegant studies of the role of microbial lipases and esterases which may cause us to revise our ideas on the role of micro-organisms in acne (Puhvel, 1975; Puhvel, Reisner & Sakamoto, 1975).

Matta (1974) has shown how *C. acnes* fails effectively to colonize the skin of prepubertal children. Presumably this is related to the secretion of sebum (Ramasastry et al., 1970) but it is clear that acquisition of a *C. acnes* flora precedes by a considerable period the appearance of clinical acne. Colonization by *Pityrosporon* species does not appear related to puberty however (Midgley & Noble in preparation).

# Intrinsic Host Factors

It is the study of the human host which has lagged most in this eco-system but there are pointers to suggest that the host would repay investigation. Hoeksma and Winkler (1963) found the nasal flora of monovular twins to be alike more often than that of single sex, binovular twins though Aly and his colleagues (1974) found less pronounced similarities. A family susceptibility to nasal carriage of *Staph. aureus* was reported by Noble et al., (1967). Racial differences have also been noted. Noble (1974) reported that negroid children less often had nasal *Staph. aureus* than caucasoid children living in the same area. Matta (1974) found negroid children to acquire *C. acnes* on the face earlier than caucasoids in the same temperate area. Nzanzumuhire et al., (1975) found, negroids in tropical areas carried more organisms on the face than children of Indian origin. D. C. Turk (personal com-munication) found that caucasoid children in Jamaica carried *Haemo-philus influenzae* more often than did negroids, whilst the carriage rate in children of Chinese origin was almost nil. In studies on fungal infec-tion Allen and his colleagues (1972) reported that negroid and caucasoid American servicemen in Vietnam had inflammatory infection due to *Trichophyton mentaarophytes* whilst Vietnamese servicemen had chronic infection due to *T. rubrum*. Blank et al., (1974) found negroid children but caucasoid adults were the more susceptible to fungus infection. Whilst social and economic factors account for some of these differences there seems little doubt that intrinsic racial differences occur.

# The Normal Skin Flora

There are immense differences in the density of skin populations of microbes. The geometric mean quantitative count of aerobic organisms ranged from $3 \cdot 1 \times 10^2$ on the calf to $1 \cdot 1 \times 10^4$ on the forehead in a series of 22 individuals examined by Noble and Somerville (1974). The individual variation was considerable however, from less than 7 colony forming units to 15 000 on the calf and from 26–2 900 000 on the forehead (Fig. 8.1). In general, males have higher counts of aerobes at most skin sites than do females although the counts of anaerobic diphtheroids do not differ significantly.

This picture of regional differences in total count is made more complex by the fact that the organisms are not spread evenly over the microlocalities. Skin micro-organisms occur in microcolonies on the

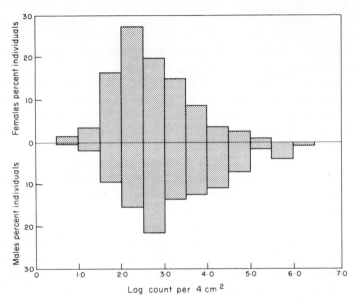

**Fig. 8.1.** *Carriage of aerobic micro-organisms on the skin of healthy adults.*

skin surface. These microcolonies are larger in some areas (e.g., fore-head) than others (e.g., forearm). On the forehead males were found to have larger microcolonies of aerobes, but smaller microcolonies of anaerobes than females (Somerville & Noble, 1973). The microcolony size of aerobes round the umbilicus is shown in Fig. 8.2.

A closer study of these microcolonies may help to clarify the concept of resident and transient flora's developed by Price (1938). Price determined that some organisms which he called the *residents*, were reproducing and maintaining themselves on the skin surface whilst others, the *transients*, were carried passively, they were purely con-taminants. That this concept was an over-simplification was recog-nized by Price himself. Recent studies on the skin flora of two individuals for 7 months have shown, as might be expected, that some organisms were present much, or all, of the time and that others appeared once only. There was however a third category, for which the term *nomads* has been proposed (Somerville-Millar & Noble, 1974) which are present, and presumably reproducing, for short periods but which do not persist indefinitely. A statistical approach may clarify some of this con-fusion since organisms which are unequivocally resident have a standard deviation which is of the order of only 25 per cent of the mean count. Transients have a greater SD which may even be equal to the mean count. Evans (1975) has concluded that the density of the microbial population is a stable individual characteristic, at least on the forehead.

An increase in humidity will increase the size of the microcolonies. The action of washing with plain or with anti-bacterial soaps seems to be to reduce the size, rather than the number of the microcolonies (Wilson, 1970). The apparent anomaly, that shower bathing increases the dispersal of organisms even when vast numbers have been removed by the shower may be explained by smearing of the microcolonies on the skin surface thus contaminating more scales than is usual (Holt, 1971). Under normal circumstances only 5–10 per cent of squames dispersed from the body carry viable bacteria.

In the aetiology of some microbial skin diseases it seems likely that the pathogens must first reach some critical population level before clinically apparent damage can occur. Dudding and his colleagues (1970) have shown how colonization of the skin precedes impetigo or streptococcal pyoderma in American Indian populations; about ten days elapsed between colonization and infection (Ferrieri et al., 1972), this may represent the time needed for the microcolonies to reach the critical size.

The normal flora is composed of members of the micrococcaceae and the diphtheroids and these are present over all the body surface. Quite striking differences in skin flora might be anticipated by comparison with the mouth where, even within an environment bathed with a common fluid (saliva) there remain differences between the flora of the tongue, tooth surface and gingivae, there are even differences in the bacterial composition of dental plaque at different sites on a tooth (Hardie & Bowden, 1974). By comparison the skin exhibits a less varied flora, perhaps because the drier environment restricts the species able to maintain themselves. There are some regional differences in bacterial distribution however, *Staph. aureus* may be isolated from the nose of about 35 per cent of both males and females and from the perineum in about 5–10 per cent, though some workers report males more often colonized at the latter site than females. Axillary carriage may also reach 5–10 per cent but *Staph. aureus* is rare, at least as a resident organism, on normal undifferentiated skin. By contrast patients with skin disease such as eczema, psoriasis or pityriasis rosea may be heavily colonized with resident *Staph. aureus* on affected and unaffected skin. It may be that studies in food science may assist in medical work for Rosky and Hamdy (1975) have identified some of the factors relating to persistence of *Staph. aureus* on meat. These include oligonucleotides which are

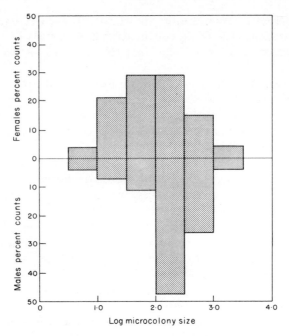

**Fig. 8.2.** *Microcolony size of the flora of the periumbilicus of normal healthy adults.*

known to have important effects on micro-organisms (Braun & Firshein, 1967).

Other difference also occur amongst the micrococcaceae, though they are less pronounced (perhaps simply less well investigated) than *Staph. aureus.* The species frequently referred to as *Sarcina* (aerobic packet-forming, Gram positive cocci with a pronounced yellow pigment) are more common on normal skin than on the specialized skin of the anterior nares or axilla. There is also an age difference, *Sarcina* are more common on children than adults.

Baird-Parker's (1965, 1974) scheme has proved of great value in identifying members of the Micrococcaceae and has identified some minor pathogens. Thus the SII strains (*Staph. epidermidis* biotype 1) which are most common in acne lesions together with the anaerobic *C. acnes* group, are also those which are the most frequent in surgical lesions, in blood stream infections and in colonization/infection of surgical prostheses. Alternative schemes have been proposed recently by Kloos and his colleagues (Kloos et al., 1974; Schleifer & Kloos,

1975; Kloos & Schleifer, 1975; Kloos et al., 1976). Bacteriophages are now available for all genera of cocci (Peters & Pulverer, 1975).

There is as yet no comparable scheme available for the 'diphtheroids'. This rather imprecise term is used advisedly since it is now apparent that genera other than the classical *Corynebacterium* may be isolated from human skin; though which are maintaining themselves as residents is less clear. D. G. Pitcher (personal communication) has identified seven major groups of aerobic 'diphtheroids' on human skin. The most frequently isolated group (62 per cent of strains) possess a typical Corynebacterium cell wall (meso D.A.P. arabino-galactan) though this type of wall is also possessed by *Nocardia* and *Mycobacteria*. These can be separated by a study of the very long chain fatty acids (mycolic acids) described by Minnikin et al., (1975). Other groups appear typical of strains previously described as plant or soil organisms, one group resembles *Propionibacterium* but is fully aerobic and one group appears not to have been described previously. It is clear that a fundamental approach of this nature may clarify many previous anomalies in taxonomic schemes for the cutaneous diphtheroids and may make studies of adjuvant diseases such as rheumatoid arthritis more meaningful in microbiological terms.

The anaerobic diphtheroids originally classed as *Corynebacterium (Propionibacterium) acnes* are now divisible into three species (Marples & McGinley, 1974) and there are suggestions that the three types may have different roles in the causation of clinically apparent acne. These organisms are members of the normal skin flora in those not suffering from acne. Those aerobic organisms which produce extracellular coproporphyrin III, and in consequence fluoresce under u.v. light, are isolated mainly from the axillae, groins and toewebs and are also implicated in the disease erythrasma, in which the skin lesions also fluoresce. Such organisms are known under the term *C. minutissimum* although many other taxonomically valid species of Corynebacterium also excrete porphyrins. Also equivocal taxonomically are the strains known as *C. tenuis* which are isolated from trichomycosis axillaris, an affection in which visible colonies of diphtheroids form on the axillary hairs. Pubic hairs are also occasionally affected. More than one biochemical type of organism is involved although some are thought to damage the hair by keratinolytic enzymes.

Thus, especially amongst the 'diphtheroids', the organisms implicated in minor skin disease are also members of the normal skin flora. They do not appear to cause harm except in those individuals who, as a result of intrinsic or extrinsic factors, become susceptible to local invasion or to gross colonization by a particular organism.

Suppression of the host immune response by micro-organisms has been admirably reviewed by Schwab (1975). Since *C. parvum*, which is being extensively studied (e.g., Scott 1974, 1975) is likely to be proven a member of the normal skin flora and to be related to *C. acnes*, studies on acne in relation to immune response would be valuable.

Gram-negative bacilli are rare on normal skin, even comparatively rare in skin disease, except for the Mimeae tribe which are resident in the axillae, groins and toe webs of perhaps 20 per cent of the normal population. Although the Gram-negative bacilli appear not to survive desiccation on the skin, there are probably always a few present, for occlusion of the surface can produce an overgrowth of Gram-negative bacilli within a few days.

## Interactions at the Skin Surface

Surprisingly little is known of the interactions of bacteria at the skin surface. Indeed less is known of skin, the most accessible organ, as a habitat, than is known of microbial interaction in the gut or in the mouth (*see* Skinner & Carr, 1974). In the mouth there have been several studies on the mechanism of adhesion of streptococci to the tooth surface to form dental plaque (e.g., Gibbons & van Houte 1971; Mukasa & Slade, 1974). Presumably bacteria on the skin are held by the surface tension of sweat and sebum, though a slightly alkaline buffer is better at removing bacteria from the skin than a neutral or acid buffer (e.g., Williamson, 1965). In the gut the chemistry of utilization of bile salts has been studied (Draser & Hill, 1974) selected recolonization of gnotobiotic mice has been carried out (Waaij et al., 1973) and the gut flora of recolonized mice has been compared with that of the skin (Waaij & Sturm, 1971). Such studies do not appear to have been extended to human skin.

In man interference has been studied most extensively in relation to nasal colonization with *Staph. aureus* strain 502A and these studies have related to the incidence of pyogenic skin lesions in neonates and adults (Maibach et al., 1969). Wickman (1970) seems to have been the only worker to try using a coagulase negative (*Staph. epidermidis*) strain for interference studies in experimental skin burns (if we except the extensive studies of McCabe (1965) in chick embryo).

Studies of the Gram-negative flora of human skin have suggested that the Gram-positive moiety normally keeps the Gram-negatives in check. Although we should not expect the general skin surface to harbour many coliform-like organisms owing to their susceptibility to desiccation, the axilla, a warm and moist area might be expected to do so. Shehadeh and Kligman (1963) found that, when the Gram-positive flora was suppressed by neomycin-containing deodorants, there was a prompt shift to a coliform flora. Use of tetracycline in acne may, in a few patients, lead to a Gram-negative folliculitis (Leyden et al., 1973).

In the laboratory, stimulation and inhibition of some skin bacteria by others may easily be demonstrated and indeed is frequently seen in routine skin cultures. It has not been demonstrated to occur in vivo however except in frank skin lesions (Selwyn, 1975). The selection of penicillin resistant *Staph. aureus* by the presence of dermatophyte fungi has been described (Smith & Marples, 1964; Wallerström, 1968) but there is reason to believe that this is not always the case. Allen and Taplin (1974) and Clayton and Noble (unpublished observations) have failed to find any correlation between the presence of dermatophytes and antibiotic resistant organisms, though production of penicillin-type substances by dermatophytes is well established (Uri et al., 1963).

Studies by Murphy (1975) have shown that although the number of chemical substances present at the skin surface may be considerable this is not sufficient to permit microbial growth except for strains of *Sarcina*. Human skin contains an accessory growth factor, as yet un-identified, which permits the growth of most skin organisms and this can be approximated to by Coenzyme A (which is not the accessory factor itself however). *Sarcina* is a very good producer of coenzyme A and releases it into the growth medium. Many skin organisms are able to grow on the defined medium when *Sarcina* is present.

The complexity of interactions between bacteria should make us wary of accepting simple hypotheses however. Haines and Harmon (1973) have shown how the inhibition of *Staph. aureus* strains by Pedio-coccus or Streptococcus may be temperature dependent. Raibaud and his colleagues (1972) have shown that, in the rat gut *Acuformis* (Clostri-dium) strains can establish only if *Ristella* (Bacteroides) is already present and that this is affected by diet.

There is a tendency to regard the skin as a static environment in which a few species of the more robust kinds of organisms are able to survive and reproduce. This picture may be far from the truth however, Major changes in the skin flora occur at puberty. Matta (1974) has shown how the skin of the forehead becomes colonized with *C. acnes* during the period of maturation. Glass (1973) has shown the prevalence of *Sarcina* to decrease with increasing age, as does that of *Staph. aureus* in the anterior nares (Noble et al., 1967).

Minor changes in flora may be brought about by the changing conditions at the skin surface, though as yet such changes can only be surmised. Studies which might be included under the term 'ecological genetics' are rare. Lacey (1971) described how resistance to neomycin might be transferred at high frequency from one strain of *Staph. aureus* to another on the skin surface. We may speculate that phage mediated transfer might occur between *Staph. aureus* and coagulase negative Gram-positive cocci, since phages exist which can act across the spectrum of cocci.

Noble (1972) has described how skin patients may carry variants of *Staph. aureus* differing in antibiotic sensitivity on the skin, these variants apparently being selected on the skin by some local ecological pressure, the nature of which is not clear. Schlaefler (1971, 1972) has found that some *Staph. epidermidis* strains which possess a penicillinase plasmid are faster growing and have simpler growth requirements than strains which lack the plasmid. In these strains the penicillinase plasmid also carried information governing phage absorption, synthesis of phospho-$\beta$-glucosidase, at least one step in the catabolism of ribose, and uptake of mannitol and biosynthesis of an unknown but essential metabolite. Clearly ecologic pressures other than the presence of penicillin may contribute to maintenance of the penicillinase-plasmid. Grinstead and Lacey (1973), noting the heavy pigmentation of strains of *Staph. aureus* isolated from skin, have shown how the paler variants are less resistant to drying and to linoleic acid than the deeply pigmented parent strains. This suggests other ecological factors which may bring about changes in the skin flora. Noble, in a continuation of the previous study, has found ecological evidence to support Grinstead and Lacey's observation. Table 8.1 shows that patients with antibiotic resistant

**Table 8.1.** *Presence of antibiotic resistant variants on patients with skin disease*

| Patient | A | | B | | C | | D | |
|---|---|---|---|---|---|---|---|---|
| | Percent of variants | | | | | | | |
| | R/P | S/P | R/P | S/P | R/T | S/T | R/T | S/T |
| Site | | | | | | | | |
| Nose | 14 | 86 | 10 | 90 | 10 | 90 | 42 | 58 |
| Skin of chest or legs | 89 | 11 | 95 | 5 | 84 | 16 | 92 | 8 |

The variants are all of the same phage type R/P = resistance to penicillin; S/P = sensitivity to penicillin; R/T = resistance to tetracycline, S/T = sensitivity to tetracycline. All patterns tested to discs of 10 units of penicillin or 30 mgm of tetracycline. Each patient had a different phage type of staphylococcus.

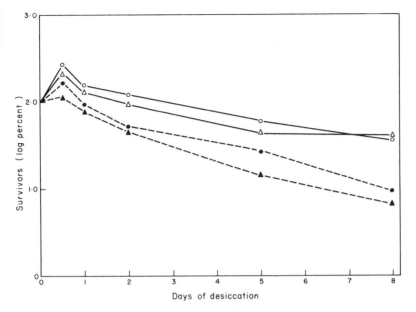

**Fig. 3.** *Survival of* Staphylococcus aureus *variants during desiccation on porcine epithelium in vitro.*
—O— Resistant to penicillin and tetracycline on plain skin
—△— Resistant to penicillin and tetracycline on skin plus sebum
--●-- Resistant to penicillin, sensitive to tetracycline on plain skin
--▲-- Resistant to penicillin, sensitive to tetracycline on skin plus sebum

variants tended to carry the more resistant organisms on unspecialized skin, such as that of the chest and legs, whilst the antibiotic sensitive strains appeared more often in the nose, an area which is both warmer and moister than the skin. In a study of the survival of these variants on commercially available sterile pig skin, washed resistant and sensitive cells were inoculated in Murphy's sweat medium on the skin in the presence or absence of the ether soluble fraction of human sebum. Survival was better in the absence of sebum and antibiotic resistant variants survived better than sensitive strains, confirming Grinstead and Lacey's observations (Fig. 8.3).

It is not known which factors may influence the creation and selection of variants, the compound dithranol (trihydroxy anthracene) much used in the treatment of psoriasis, will induce petite mutants of yeasts (Swanbeck & Thyresson, 1974) and we may speculate that other compounds applied to human skin will also influence the flora. Some antibiotics, such as the tetracyclines, appear at the skin surface following oral absorption while others, such as the penicillins, fail to appear in more than minute amounts (Marples & Kligman, 1971).

There may be much to be learned from other, surprisingly similar habitats. In a symposium on leaf surface organisms (Preece & Dickinson, 1971) descriptions may be found of the excretion of amino acids and sugars on the leaf surface, of the growth of pathogens and non-pathogens and how fungal penetration of the leaf cuticle may occur despite the production and excretion of the phytoalexins, specific antifungal substances produced by the host plant. Perhaps it is in the interactions between inhabitants that this area has most to teach the medical microbiologist and it might be interesting to follow, for example, the work of Dunn and his colleagues (1971) in their demonstration that glucose, which is freely present on both plant and human surface, enables plant micro-organisms to avoid the inhibitory effects of fungicides.

If there has been a tendency in this article to stress what is unknown in the ecology of human skin, it is because there are many gaps in our knowledge of this ubiquitous and easily accessible habitat. The human skin offers great scope in all areas of microbiology; nutritional and ecological studies are required no less than investigations of the taxonomy and physiology of the cutaneous flora. A thorough understanding of the skin as a habitat will lead to a greater appreciation of the role of micro-organisms in skin disease and to the maintenance of the skin in a healthy state, as well as affording a habitat for ecologic experiment and investigation as diverse and interesting as any yet explored.

## References

Allen, A. M. & Taplin, D. (1974) Skin infections in Eastern Panama. *Am. J. trop. Med. Hyg.*, **23**, 950.

Allen, A. M., Taplin, D., Lowy, J. A. & Twigg, L. (1972) Skin infections in Vietnam. *Milit. Med.*, **137**, 295.

Aly, R., Maibach, H. I., Shinefield, H. R. & Strauss, W. G. (1972) Survival of pathogenic microorganisms on skin. *J. invest. Derm.*, **58**, 205.

Aly, R., Maibach, H. I., Shinefield, H. R. & Mandel, A. D. (1974) Staphylococcus aureus carriage in twins. *Am. J. Dis. Child.*, **127**, 486.

Aly, R., Maibach, H. I., Rahman, R., Shinefield, H. R. & Mandel, A. D. (1975) Correlation of human in vivo and in vitro cutaneous antimicrobial factors. *J. infect. Dis.*, **131**, 579.

Baird-Parker, A. C. (1965) The classification of staphylococci and micrococci from world wide sources. *J. gen. Microbiol.*, **38**, 363.

Baird-Parker, A. C. (1974) The basis for the present classification of *Staphylococci* and *Micrococci*. *Ann. N.Y. Acad. Sci.*, **236**, 7.

Behrendt, H. & Green, M. (1971) *Patterns of skin pH from birth through adolescence.* Springfield: Illinois, Thomas.

Blank, F., Mann, S. J. & Reale, R. A. (1974) Distribution of dermatophytosis according to age, ethnic group and sex. *Sabouraudia,* **12**, 352.

Braun, W. & Firshein, W. (1967) Biodynamic effects of oligonucleotides. *Bact. Rev.,* **31**, 83.

Burton, J. L., Laschet, U. & Shuster, S. (1973). Reduction of sebum excretion in man by the antiandrogen, cyproterone acetate. *Br. J. Derm.,* **89**, 487.

Cabau, N., Levy, F.-M. & Muller, O. (1974) Immunoglobulins from the sweat glands of children. *Path. Microbiol.,* **22**, 883.

Chujoh, T. (1971) Studies on bacterial resident flora of the normal skin surface. *Jap. J. Derm., Series B.* **81**, 168.

Coutinho, C. B. & Nutini, L. G. (1963). Correlation between the essential amino acid requirements of *Staphylococcus aureus*, their phage types and antibiotic resistance patterns. *Nature, Lond..* **198**, 812.

Downing, D. T., Strauss, J. S., Ramasastry, P., Abel, M., Lees, C. W. & Pochi, P. E. (1975) Measurement of the time between synthesis and surface excretion of sebaceous lipids in sheep and man. *J. invest. Derm.,* **64**, 215.

Drasar, B. S. & Hill, M. J. (1974) *Human intestinal flora.* London, Academic Press.

Dudding, B. A., Burnett, J. W., Chapman, S. S. & Wannamaker, L. W. (1970) The role of the normal skin in the spread of streptococcal pyoderma. *J. Hyg., Camb.,* **68**, 19.

Duncan, W. C., McBride, M. E. & Knox, J. M. (1969) Bacterial flora. The role of environmental factors. *J. invest. Derm.,* **52**, 479.

Dunn, C. L., Benyon, K. I., Brown, K. F. & Montagne, J. T. W. (1971) The effect of glucose in leaf exudates upon the biological activity of some fungicides, p. 491. *In: Ecology of leaf surface microorganisms,* Eds T. F. Preece & C. H. Dickinson, London: Academic Press.

Eberhardt, H. (1974) The regulation of sebum excretion in man. *Arch. derm. Forsch.,* **251**, 155.

Evans, C. A. (1975) Persistent individual differences in the bacterial flora of the skin of the forehead: Numbers of Propionibacteria. *J. invest. Derm.,* **64**, 42.

Farrior, J. W. & Kloos, W. E. (1975) Amino acid and vitamin requirements of *Micrococcus* species isolated from human skin. *Int. J. systemat. Bact.,* **25**, 80.

Ferrieri, Patricia, Dajani, A. S., Wannamaker, L. W. & Chapman, S. S. (1972) Natural history of impetigo. 1. Site of acquisition and familial patterns of spread of cutaneous streptococci. *J. clin. Invest.,* **51**, 2851.

Förström, L., Goldyne, M. E. & Winkelmann, R. K. (1975) IgE in human eccrine sweat. *J. invest. Derm.,* **64**, 156.

Gibbons, R. J. & van Houte, J. (1971) Selective bacterial adherence to oral epithelial surfaces and its role as an ecological determinant. *Infect. Immun..* **3**, 567.

Glass, M. (1973) *Sarcina* species on the skin of the human forearm. *Trans. St. John's Hosp. derm. Soc.,* **59**, 56.

Gloor, M. (1975) Über die Hautoberflächenlipide. *Hautarzt.* **26**, 6.

Grinstead, J. & Lacey, R. W. (1973) Ecological and genetic implications of pigmentation in *Staphylococcus aureus. J. gen. Microbiol..* **75**, 259.

Haines, W. C. & Harman, L. G. (1973) Effect of variation in conditions of incubation upon inhibition of *Staphylococcus aureus* by *Pediococcus cerevisiae* and *Streptococcus lactis*. *Appl. Microbiol.*, **25**, 169.

Halvorsan, H. (1972) Utilization of single L-amino-acids as sole source of carbon and nitrogen by bacteria. *Can. J. Microbiol.*, **18**, 1647.

Hammar, H. (1975) Epidermal nicotinamide dinucleotides in psoriasis and neurodermatitis (lichen simplex hypertrophicus) *Arch. derm. Forsch.*, **252**, 217.

Hardie, J. M. & Bowden, G. H. (1974) The normal microbial flora of the mouth. p. 47. *In: The normal microbial flora of man*. Eds F. A. Skinner & J. G. Carr. London: Academic Press.

Hoeksma, A. & Winkler, K. C. (1963) The normal flora of the nose in twins. *Acta leidensia*, **32**, 124.

Holt, R. J. (1971) Aerobic bacterial counts on human skin after bathing. *J. med. Microbiol.*, **4**, 319.

Horne, G. O. (1952) Climate and environmental factors in the etiology of skin disease. *J. invest. Derm.*, **18**, 107.

Jenkinson, D. McK., Mabon, R. M. & Manson, W. (1974) The effect of temperature and humidity on the losses of nitrogenous substances from the skin of Ayrshire calves. *Res. vet. Sci.*, **17**, 75.

Jones, H. E., Reinhardt, J. H. & Rinaldi. M. G. (1974) Model dermatophytosis in naturally infected subjects. *Arch. Derm.*, **110**, 369.

Kellum, R. E. & Strangfeld, K. (1972) Acne vulgaris: Studies in pathogenesis. Fatty acids of human surface triglycerides from patients with and without acne. *J. invest. Derm.*, **58**, 315.

Kloos, W. E. & Schleifer, K. H. (1975) Isolation and characterization of staphylococci from human skin. *Int. J. systemat. Bact.*, **25**, 62.

Kloos, W. E., Tornabene, T. G. & Schleifer, K. H. (1974) The isolation and characterization of Micrococci from human skin, including the new species *Micrococcus lylae* and *Micrococcus kristinae*. *Int. J. systemat. Bacteriol.*, **24**, 79.

Kloos, W. E., Scheifer, K. H. & Noble, W. C. (1976) Estimation of character parameters in coagulase negative *Staphylococcus species*. *Proceedings of the 3rd International Symposium on Staphylococci and Staphylococcal Infection*, Warsaw.

Knight, A. G. (1972) A review of experimental human fungus infections. *J. invest. Derm.*, **59**, 354.

Kuno, Y. (1956) *Human perspiration*. Springfield, Illinois: Thomas.

Lacey, R. W. (1968a) Binding of neomycin and analogues by fatty acids in vitro. *J. clin. Path.*, **21**, 564.

Lacey, R. W. (1968b) Antibacterial action of human skin. In vivo action of acetone, alcohol and soap on behaviour of *Staphylococcus aureus*. *Br. J. exp. Path.*, **49**, 209.

Lacey, R. W. (1969) Loss of antibacterial action of skin after topical neomycin. *Br. J. Derm.*, **81**, 435.

Lacey, R. W. (1971) High frequency transfer of neomycin resistance between naturally occurring strains of *Staphylococcus aureus*. *J. med. Microbiol.*, **4**, 73.

Leyden, J. J., Marples, R. R., Mills, O. H. & Kligman, A. M. (1973) Gram-negative folliculitis—A complication of antibiotic therapy in acne vulgaris. *Br. J. Derm.*, **88**, 533.

Mabon, R. M. (1974) Deoxyribonucleic acid (DNA) in cattle skin washings. *Br. J. Derm.*, **91**, 271.

Marples, Mary J. (1965) *Ecology of Human Skin*, Springfield, Illinois: Thomas.

Marples, R. R. & Kligman, A. M. (1969) Growth of bacteria under adhesive tapes. *Arch. Derm.*, **99**, 107.

Marples, R. R. & Kligman, A. M. (1971). Ecological effects of oral antibiotics on the microflora of human skin. *Arch. Derm.*, **103**, 148.

Marples, R. R. & McGinley, K. J. (1974) *Corynebacterium acnes* and other anaerobic diphtheroids from human skin. *J. med. Microbiol.*, **7**, 349.

Marples, R. R., Leyden, J. J., Stewart, Rebecca, N., Mills, O. H. & Kligman, A. M. (1974) The skin flora in acne vulgaris. *J. invest. Derm.*, **62**, 37.

Maibach, H. I., Strauss, W. G. & Shinefield, H. R. (1969) Bacterial interference: relating to chronic furunculosis in man. *Br. J. Derm.*, **81**, *Suppl. 1*, page 69.

Matta, M. (1974). Carriage of *Corynebacterium acnes* in school children in relation to age and race. *Br. J. Derm.*, **91**, 557.

McBride, M. E., Duncan, W. C. & Knox, J. M. (1975) Physiological and environmental control of Gram-negative bacteria on skin. *Br. J. Derm.*, **93**, 191.

McCabe, W. R. (1965) Staphylococcal interference in infection in embryonated eggs. *Nature, Lond.*, **205**, 1023.

Mier, P. D. & McCabe, M. G. P. (1963) The distribution of phosphorus in the lesions of eczema. psoriasis and seborrhoeic dermatitis. *Br. J. Derm.*, **75**, 354.

Minnikin, D. E., Alshamaony, L. & Goodfellow, M. (1975). Differentiation of *Mycobacterium, Nocardia* and related taxa by thin layer chromatographic analysis of whole organism methanolysates. *J. gen. Microgiol.*, **88**, 200.

Molin, L. & Wester, P. O. (1973a) Iron content in normal and psoriatic epidermis. *Acta derm.-vener.*, *Stockh.*, **53**, 473.

Molin, L. & Wester, P. O. (1973b) Cobalt, copper and zinc in normal and psoriatic epidermis. *Acta derm.-vener.*, *Stockh.*, **53**, 477.

Molin, L. & Wester, P. O. (1974). Trace elements with suspected and hitherto unknown biological function in normal and psoriatic epidermis. *Acta Dermatovenereologica (Stockholm)*, **54**, 49.

Mukasa, H. & Slade, H. D. (1974) Mechanism of adherence of Streptococcus mutans to smooth surfaces. *Infect. Immun.*, **9**, 419.

Murphy, C. T. (1975) Nutrient materials and the growth of bacteria on human skin. *Trans. St. John's Hosp. derm. Soc.*, **61**, 51.

Neufahrt, Anita, Forster, J., Besser, H. & Balikcioglu, S. (1975). Isolation and amino acid composition of two pathologically augmented proteins from psoriatic scales. *Arch. derm. Forsch.*, **252**, 305.

Noble, W. C. (1972) Loss of antibiotic resistance in staphylococci from a skin hospital. *Lancet*, **1**, 929.

Noble, W. C. (1974) Carriage of *Staphylococcus aureus* and beta haemolytic streptococci in relation to race. *Acta derm.-vener.*, *Stockh.*, **54**, 403.

Noble, W. C. (1975a) Skin as a microhabitat. *Post-grad. med. J.*, **51**, 151.

Noble, W. C. (1975b) Dispersal of skin microorganisms. *Br. J. Derm.*, **93**, 477.

Noble, W. C. & Somerville, D. A. (1974) *Microbiology of human skin*, London: W. B. Saunders.

Noble, W. C., Valkenburg, H. A. & Wolters, C. H. L. (1967) Carriage of *Staphylococcus aureus* in random samples of a normal population. *J. Hyg., Camb.*, **65**, 567.

Nzanzumuhire, H., Masawe, A. J. E. & Mhalu, F. S. (1975) The bacteriological ecosystem of the skin of children in an African tropical environment (Tanzania). *Br. J. Derm.*, **92**, 77.

Pablo, G., Hammons, A., Bradley, S. & Fulton, J. E. (1974) Characteristics of the extracellular lipases from *Corynebacterium acnes* and *Staphylococcus epidermidis. J. invest. Derm.*, **63**, 231.

Pablo, G. M. & Fulton, J. E. (1975) Sebum. Analysis by infra-red spectroscopy. II. The suppression of fatty acids by systematically administered antibiotics. *Archs Derm.*, **111**, 734.

Payne, A. M. M. (1949) The influence of humidity on the survival of *Bacterium coli* on the skin. *Mon. Bull. Minist. Hlth*, **8**, 263.

Peters, G. & Pulverer, G. (1975) Bacteriophages from micrococci. *Zentbl. Bakt. Hyg., I Abt., Originele A*, **232**, 221.

Pillsbury, D. M. & Rebell, G. (1952) The bacterial flora of the skin. Factors influencing the growth of resident and transient organisms. *J. invest. Derm.*, **18**, 173.

Plewig, G. (1970) Regional differences in cell sizes in the human stratum corneum. *J. invest. Derm.*, **54**, 19.

Pollock, M. R., Wainwright, S. D. & Mansion, E. E. D. (1949) The presence of oleic acid requiring diphtheroids on human skin. *J. Path. Bact.*, **61**, 274.

Preece, T. F. & Dickinson, C. H. (Eds) (1971). *Ecology of leaf surface microorganisms.* London: Academic Press.

Price, P. B. (1938) The bacteriology of normal skin: A new quantitative test applied to a study of the bacterial flora and the disinfectant action of mechanical cleansing. *J. infect. Dis.*, **63**, 301.

Puhvel, S. Madli (1975) Esterification of [4-$^{14}$C] cholesterol by cutaneous bacteria (*Staphylococcus epidermidis, Propionibacterium acnes* and *Propionibacterium granulosum*). *J. invest. Derm.*, **64**, 397.

Puhvel, S. Madli, Reisner, R. M. & Sakamoto, M. (1975) Analysis of lipid composition of isolated human sebaceous gland homogenates after incubation with cutaneous bacteria. Thin layer chromatography. *J. invest. Derm.*, **64**, 406.

Puhvel, S. M. & Reisner, R. M. (1970) Effect of fatty acids on the growth of *Corynebacterium acnes* in vitro. *J. invest. Derm.*, **54**, 48.

Raibaud, P., Ducluzeau, R., Muller, M. C. & Abrams, G. D. (1972) Diet and the equilibrium between bacteria and yeast implanted in gnotobiotic rats. *Am. J. clin. Nutr.*, **25**, 1467.

Ramasastry, P., Downing, D. T., Pochi, P. E. & Strauss, J. S. (1970) Chemical composition of human skin surface lipids from birth to puberty. *J. invest. Derm.*, **54**, 139.

Rosky, C. T. & Hamdy, M. K. (1975) Persistence of staphylococcus. Bruised tissue microenvironment affecting persistence of *Staphylococcus aureus. J. Food Sci.*, **40**, 496.

Savin, J. A. (1974) Bacterial infections in diabetes mellitus. *Br. J. Derm.*, **91**, 481.

Savin, J. A. & Noble, W. C. (1975) Immunosuppression and skin infection. *Br. J. Derm.*, **93**, 115.

Schaefler, S. (1971) *Staphylococcus epidermidis BV.* Antibiotic resistance patterns, physiological characteristics and bacteriophage susceptibility. *Appl. Microbiol.*, **22**, 693.

Schaefler, S. (1972) Polyfunctional penicillinase plasmid in *Staphylococcus epidermidis.* Bacteriophage restriction and modification mutants. *J. Bact.*, **112**, 697.

Schleifer, K. H. & Kloos, W. E. (1975) Isolation and characterization of staphylococci from human skin. *Int. J. systemat. Bact.,* **25**, 50.

Schwab, J. H. (1975) Suppression of the immune response by microorganisms. *Bact. Rev.,* **39**, 121.

Scott, M. T. (1974) *Corynebacterium parvum* as a therapeutic anti-cancer agent. *Seminars in Oncology,* **1**, 367.

Scott, M. T. (1975) In vivo cortizone sensitivity of non specific antitumor activity of *Corynebacterium parvum*—activated mouse peritoneal macrophages. *J. Nat. Cancer Inst.,* **54**, 789.

Selwyn, S. (1975) Natural antibiosis among skin bacteria as a primary defence against infection. *Br. J. Derm.,* **93**, 481.

Shehadeh, N. H. & Kligman, A. M. (1963) The effect of topical antibacterial agents on the bacterial flora of the axilla. *J. Invest. Derm.,* **40**, 61.

Shifrine, M. & Marr, A. G. (1963) The requirement for fatty acids by *Pityrosporum ovale. J. gen. Microbiol.,* **32**, 263.

Singh, G. (1973) Experimental trichophyton infection of intact human skin. *Br. J. Derm.,* **89**, 595.

Singh, G. (1974) Heat, humidity and pyodermas. *Dermatologica,* **147**, 342.

Skinner, F. A. & Carr, J. G. (eds) (1974) *The Normal Microbial Flora of Man.* London: Academic Press.

Smith, J. M. B. & Marples, M. J. (1964) A natural reservoir of penicillin resistant strains of *Staphylococcus aureus. Nature, Lond.,* **201**, 844.

Smith, R. F. (1969) Characterization of human cutaneous diphtheroids. *J. gen. Microbiol.,* **55**, 433.

Smith, R. F. (1971) Lactic acid utilization by the cutaneous *Micrococcaceae. Appl. Microbiol.,* **21**, 777.

Smith, R. F. & Willett, N. P. (1968) Lipolytic activity of human cutaneous bacteria. *J. gen. Microbiol.,* **52**, 441.

Sobhanadri, C., Ramamurthy, P. J. & Reddy, M. A. (1974) Free amino acid pattern of psoriatic scale. *Indian J. Derm. Vener.,* **40**, 135.

Somerville, D. A. (1973) A taxonomic scheme for aerobic diphtheroids from human skin. *J. med. Microbiol.,* **6**, 215.

Somerville, D. A. & Lancaster-Smith, M. (1973) The aerobic cutaneous microflora of diabetic subjects. *Br. J. Derm.,* **89**, 395.

Somerville, D. A. & Noble, W. C. (1973) Microcolony size of microbes on human skin. *J. med. Microbiol.,* **6**, 323.

Somerville-Millar, D. A. & Noble, W. C. (1974) Resident and transient bacteria of the skin. *J. cutan. Path.,* **1**, 260.

Swanbeck, G. & Thyresson, M. (1974) Induction of respiration deficient mutants in yeast by psoralen and light. *J. invest. Derm.,* **63**, 242.

Taplin, D., Lansdell, Lyle, Allen, A. M., Rodriguez, R. & Corton, A. (1973) Prevalence of streptococcal pyoderma in relation to climate and hygiene, *Lancet,* **1**, 501.

Taplin, D. & Allen, A. M. (1974) Bacterial pyodermas. *Clin. Pharm. Thera.,* **16**, 905.

Taplin, D., Zaias, N. & Rebell, G. (1965) Environmental influences on the microbiology of the skin. *Arch. envir. Hlth,* **11**, 546.

Uri, J., Valu, G. & Bekesi, I. (1963) Production of 6-aminopenicillanic acid by dermatophytes. *Nature, Lond.,* **200**, 896.

Waaji, D. van der & Sturm, C. A. (1971) The production of 'bacteria free' mice. Relationship between faecel flora and bacterial populations of the skin. *Antonie van Leeuwenhoek,* **37,** 139.

Waaij, D. van der, Vossen, J. M., Kal, H. B. & Speltie, T. M. (1973) Bio-typing of Enterobacteriaceae—an important tool in the evaluation of systems for protective isolation. p. 546, *In: Airborne Transmission and Airborne Infection,* Eds J. F. Ph. Hers & K. C. Winkler. Utrecht: Oosthoek.

Wallerström, A. (1968) Production of antibiotics by *Epidermophyton floccosum.* II. Microflora in *Epidermophyton* infected skin and its resistance to antibiotics produced by the fungus. *Acta. path. microbiol. scand.,* **74,** 531.

Weeks, G., Shapiro, M., Burns, R. O. & Wakil, S. J. (1969) Control of fatty acid metabolism. 1 Induction of the enzymes of fatty acid oxidation in *Escherichia coli. J. Bact.,* **97,** 827.

Weirich, E. G. & Longauer, J. (1974) Inhibition of sebaceous glands by topical application of oestrogen and anti-androgen on the auricular skin of rabbits. *Arch. derm. Forsch.,* **250,** 81.

Wickman, K. (1970) Studies of bacterial interference in experimentally produced burns in guinea-pigs. *Acta path. microbiol. scand.,* **B, 78,** 15.

Williamson, P. (1965) Quantitative estimation of cutaneous bacteria. p. 3, *In: Skin Bacteria and their Role in Infection,* Eds H. I. Maibach & G. Hildick-Smith. New York: McGraw-Hill.

Wilson, P. E. (1970) A comparison of methods for assessing the value of antibacterial soaps. *J. appl. Bact.,* **33,** 574.

Zaun, H. & El Mozayan, M. (1973) Hautoberflächenzucker als Milieufaktor für die Mikrobielle Besiedlung der Haut. *Hautarzt,* **24,** 428.

# Index

257